CHANGING PLAC

EN

CHANGING PLACES

WOMEN'S LIVES IN THE CITY

edited by
Chris Booth
Jane Darke
and
Susan Yeandle

P·C·P
Paul Chapman
Publishing Ltd

Paul Chapman Publishing Ltd
144 Liverpool Road
London N1 1LA

British Library Cataloguing in Publication Data
Changing places : women's lives in the city
1. Sociology, Urban 2. Women – Social conditions
I. Booth, Chris II. Darke, Jane III. Yeandle, Susan
307.7'6'082

ISBN 1 85396 311 9

Typeset by Anneset, Weston-super-Mare, North Somerset
Printed and bound in Great Britain

ABCD9876

CONTENTS

PREFACE

Women's Lives in the City was first conceived in 1993 when two of the authors developed a new undergraduate module on Women and Cities for students of urban studies, planning and housing at Sheffield Hallam University. At that time the available literature in this field appeared fragmented and partial. Consequently, the book aimed to provide a suitable text to support this module, which would also be useful to students of women's studies wishing to analyse the built environment. Although other related texts have been published recently, the present text offers an original and important contribution as it provides a critical feminist commentary on women's experiences of living in and using the urban environment. The editors have brought together authors from a variety of backgrounds: town planning, transport, housing, architecture and social sciences. The book stresses the multiplicity of ways in which women experience urban life, discussing positive aspects of the choices that some women now enjoy as well as the experience of disadvantage and oppression. The diversity of the authors' backgrounds and experience gives the text a richness and depth.

The draft chapters have all been 'field tested' on final year undergraduate students in the School of Urban and Regional Studies at Sheffield Hallam University. The authors have found that not only did students greatly enjoy reading the text, but it also helped change their perceptions of women's experience in living in the city. The drafts stimulated lively discussion and provided a useful platform from which to encourage students to add breadth and depth to their reading in this field. We would like to thank these students and our colleagues for their constructive comments.

A word about language. Many books written from a feminist perspective use the first person plural to refer to a presumed shared experience encompassing the writer(s) and women readers. The authors have decided generally to avoid this usage. Our reasons are, firstly, that it excludes male readers (in the same way that women have often been excluded in the past by the use of male pronouns to refer to a male norm applying to both sexes) whereas we hope that men will read and think about this book. Secondly, this usage elides the *differences* between women which is one of the themes of this book. Thirdly, it is presumptuous of the authors to claim a universality for our thoughts and experiences. Personal experience is valuable but by definition nobody can fully share the experiences of another, so *empathy* is equally important: empathy across the social divisions of gender, ethnicity, age, income level and so on.

The editors wish to thank Marianne Lagrange of Paul Chapman publishers for her support, and the Centre for Regional Economic and Social Research (CRESR) at Sheffield Hallam University for providing space and time out from regular teaching duties. Finally, warmest thanks are due to the supportive others in the authors' households.

Chris Booth, Jane Darke, Sue Yeandle
1996

NOTES ON CONTRIBUTORS

Chris Booth is a Senior Lecturer in Planning Practice at Sheffield Hallam University. For over twenty years she has worked for a variety of public and private sector organisations, specialising in consultation and client-based approaches to planning. She is co-author of the RTPI's Practice Advice Note 12, *Planning for women*.

Jane Darke is Senior Lecturer in Housing and Equal Opportunities in the School of Planning at Oxford Brookes University. Previously she worked in local authority housing departments, then taught at Sheffield Hallam University. She has written extensively on housing.

Eileen Green is Professor of Sociology at the University of Teesside where she is Director of the Centre for the Study of Adult Life. Previously she taught sociology at Sheffield Hallam University. Her research interests are in women and leisure, women and the labour market, and gender, organisational change and IT. She has published extensively in these areas.

Rosalie Hill has recently retired from her post as Senior Lecturer in Planning at Sheffield Hallam University to devote more time to research. She has considerable experience of both transport data and policy analysis, and has special interests in the transport needs and behaviour of different social groups, including women. She has published and broadcast extensively, with particular reference to the South Yorkshire Policy context.

Penny Lidstone is completing a PhD in the Centre for Regional Economic and Social Research at Sheffield Hallam University. Her research publications include 'Rationing Housing to the Homeless Applicant' (*Housing Studies*) and her thesis is concerned with the role of administrative discretion in the allocation of housing to homeless applicants.

Helen Morrell is a Senior Research Officer at the Department for Education and Employment. She previously worked at the Centre for Regional Economic and Social Research at Sheffield Hallam University, where she undertook an evaluation of a women's safety strategy funded by the Home Office *Safer Cities* programme.

Dory Reeves lectures in the Department of Environmental Planning at the University of Strathclyde, and is a member of the Royal Town Planning Institute (RTPI)'s Equal Opportunities (Women) Panel. With over ten years' experience in housing and planning in local government and research, she has specialised in housing, development planning, public consultation, management for planners and planning aid.

Carol Walker is Principal Lecturer in Social Policy at Sheffield Hallam University. She has undertaken research on various aspects of social security. Her key publications in this area include *Managing Poverty: the limits of social assistance* (Routledge, 1993). She has also undertaken research and published widely on service provision for people with learning difficulties living in the community.

Susan Yeandle is Principal Lecturer in Sociology at Sheffield Hallam University, where she also co-ordinates the Women's Studies Group in the Centre for Regional Economic and Social Research. She has written extensively on women and the labour market, and has a particular interest in the connections between work and family life.

INTRODUCTION

1.

WOMEN, FEMINISMS AND METHODS

Susan Yeandle

CENTRAL AND UNIFYING ASPECTS OF FEMINIST ANALYSIS

While feminist approaches involve using a diverse range of theories, methods and empirical material, most feminists would agree that there are certain key aspects of feminist thinking which they share. This book offers a specifically feminist approach to the topic of women and cities and in the following chapters, readers will see that its contributors offer a variety of ways of approaching their subjects. In this opening chapter we want to outline what unifies feminist approaches, as well as indicating the wide range of ways in which they may differ.

Above all, feminist academic work seeks to understand a phenomenon which all feminists recognise, although they may give it different names: women's disadvantage, subordination or oppression.[1] Feminist work seeks to explain women's subordination through careful analysis: in turn, that analysis forms part of a broader aim of finding ways of overcoming women's subordination. This means, as we shall see below, that feminist academic and theoretical work should normally be understood as a guide to action and to politics, rather than as an end in itself.

While there is considerable disagreement among feminists about the precise cause(s) of women's subordination, feminists share a recognition that, historically, women's biology, notably in relation to pregnancy, childbirth and lactation, has meant they have certain common experiences and common needs. Consequently women have come to claim certain rights which also differentiate them from men. This biological aspect of women's situation — unless and until it is overcome — has also meant that women are subjected to particular forms of exploitation which separate them from men. Furthermore, women's sexual availability to men has often been assumed, and men have frequently asserted a right to sexual gratification through access to women's bodies. It is therefore no accident that campaigns for birth control and maternity services, for protection from sexually transmitted diseases, and for a woman's right to choose in relation to abortion and

2

childbirth, have featured prominently in the history of feminist political activity — alongside the other well-known campaigns for women's right to education, suffrage, employment opportunities and for legal rights within marriage.

In trying to understand women's situation and women's subordination in contemporary cities, we find that women's places, women's experiences and women's activities today still separate them from men to a rather marked degree. This is not to claim that no contemporary urban experiences are shared between men and women, but rather to note that in their homes and home-based activities, in their ways of moving about cities and of using transport systems, in their activities and participation in urban communities, in their caring and support for family members, in their work for pay or 'for free', in their use of urban services and of shops, women often have more in common with other women than they do with men.

This unity of experience needs to be set against the divisions between women which are discussed below and elsewhere in this book. It is for the reader to judge how far women's unity of experience is a measure of subordination or of strength, and whether it offers a possible basis for women to change their situation or rather renders them continually disempowered, underlining the entrenched nature of the inequalities against which women have been struggling for more than 200 years.

In this book, we want to join with other feminists in turning the focus of academic, policy and theoretical work on urban questions on to women's experience and potential. Part of the task of women's studies is to draw attention to the way in which much theorising and policy-making has used men's activities and experiences as a norm, often hiding women's lives from view. Our gaze in this book is on those female lives. In it, we develop our thesis that both those experiences and characteristics which women share and those which differentiate them require analysis and explanation: women's lives in the city — in all modern urban environments — are varied, alter as they age and move through their lifecourse and, further, that they may be affected by the special privileges or disadvantages of their class, race, religion and (dis)ability of body or mind.

DIVERSIFICATION: DIFFERENT FEMINIST APPROACHES — THEIR ORIGINS, THEORIES AND CONSEQUENCES

Although these unifying aspects of feminist analysis form its core, feminist theories and feminist practices have become more diverse and complex as feminists have grappled with changing historical circumstances and have drawn upon and developed a variety of theoretical traditions. Tong (1992) offers one of a number of comprehensive introductions to feminist thought which explores dimensions of feminist theorising which can only be touched on in this introduction.

Liberal feminism

The tradition of liberal feminism, in which recourse is made to concepts of reason, law, freedom and rights — and by extension to questions of discrimination, fairness and equality of opportunity — emerges from the nineteenth-century classical liberal body of thought associated with J.S. Mill and de Tocqueville, with its focus above all on the individual and on individualism.

Liberal feminism has roots in the revolutions of the late eighteenth century and in the work of Mary Wollstonecraft,[2] but in Britain became an especially important focus for thoughtful women and some men in the mid to late nineteenth century, drawing into its orbit campaigns for women's family and property rights, for women's suffrage and for women's access to education and the professions. In the twentieth century it has been associated with analyses of women's disadvantage and with claims for equality of opportunity, and with the development of strategies to overcome those disadvantages, often using legislation as the vehicle for change. Changes in law and in policy have been the focus of successful campaigns in a variety of fields: for equality of treatment with men in the workplace, access to education and training on a par with that available to boys and men, the establishment of fair practices in the field of services, property and taxation, and a concern to minimise the ways in which the female condition — notably with regard to pregnancy, maternity and child-rearing — should jeopardise women's chances of an equal place with men in the world.

Marxist-feminism

In Marxist-feminism the central focus is — as in classical Marxism — located in the relations of production, and consequently with class analysis. The relationship between wage labourer and capitalist 'owner of the means of production' forms the axis around which all social relations are located in Marxist thought. Marxist-feminism focuses on an analysis specifically of women's labour, and draws into that analysis elements excluded by traditional Marxist accounts. The focus is upon women's work, both in employment as workers in the usual Marxist sense, and in their activities as mothers, wives and carers, usually in the context of the heterosexual family. Marxist feminists often use the term 'social reproduction' to refer to this caring, nurturing and reproductive work, and this has led to the development of analysis of the relations of social reproduction. In this theoretical work, Marxist feminists have developed certain key Marxist concepts and have extended debate about them: the concept of women as a reserve army of labour,[3] the claim that domestic labour is expropriated from women in the interests of capital,[4] and the argument that class position divides women from each other (Rowbotham, 1973). In addition, the Marxist concept of alienation has been important in developing a Marxist-feminist account of housework, domestic labour and family relations.

Radical feminism

This approach arises from a concern with women's relationship with men, and tends to give prominence to reproduction, sexuality, bodies, male violence and coercion in analysing women's oppression. That oppression is seen by radical feminists as well-nigh universal, as a prior form of oppression — in the sense, for example, that it pre-dates capitalism — and as deep-rooted: the solutions proposed by either liberal or Marxist feminists will not be adequate to eradicate women's oppression, according to radical feminists.

Radical feminism burst into the academy in the 1960s and 1970s, although many key texts in this tradition were written long before. In the work of Millett, Daly and Rich, feminism found theorists who could attract widespread comment, and who would introduce a new generation of feminists to the role of violence, sexuality, bodies and biology in women's subordination. Rich paid attention to motherhood and to 'compulsory heterosexuality', but did so in a way which drew attention to women's power and to women's potential. She argued for social transformation rather than equality, emphasising the importance of 'the ordinary' in women: it is this : 'which will rise in every sense of the word — spiritually and in activism. . . . The "common woman" is in fact the embodiment of the extraordinary will-to-survival in millions of obscure women' (Rich, 1980). Radical feminism is also frequently, although not necessarily, connected with lesbian choice and identity — in some cases specifically identified as a political strategy of avoiding engagement in any intimacy with men.

These three strands — liberal, Marxist and radical feminist — are often separately identified in accounts of feminist theory, and certainly represent some historical and continuing divisions between feminist writers and activists. Each of these strands is explored in more detail in the chapters which follow, and within a specific context. The chapters all include references to the relevant literature as a guide to further reading.

Two further developments, both more recent in provenance, should also be noted. These are important because they offer the possibility of — at least in part — overcoming divisions between feminisms. The first involves the construction — out of Marxist and radical feminism — of 'dual systems' theory, while the second development, the emergence of a post-modernist feminism, claims that late twentieth-century societies are changing at a pace and to a degree analogous to the change wrought by the industrial revolution. It also takes issue with the category 'woman', raising important questions about women's identities and politics. Here, scientific and technological change acquires a new significance, and introduces new sources of power into the analysis of social relations.

Dual systems theory

Dual systems theorists include analyses of both capitalism and patriarchy in their explanation of women's subordination, seeing these as two systems or structures which are analytically distinct, even if empirically connected. Their approach has progressed debate in that a straight fight between alternative conceptualisations of 'the enemy' or 'the cause' of women's subordination (irrationality/men/capitalist exploitation) is avoided, and with it the divisions between women which these can engender.

Mitchell (1971) brought cultural aspects and ideology into the argument (via an exploration of psychoanalytical theory), alongside material conditions: to her, women's internalisation of patriarchy deep in the psyche was just as important as their economic and material situation; indeed, the two were mutually reinforcing and interdependent. Hartmann (1981) argued that there was a material dimension both to women's oppression by men and to their subordination within capital, while Walby (1990) has developed a more complex analysis which, some may argue, leads her beyond an analysis of dual systems. In her account, there are at least six structures which shape relations between men and women, and which are oppressive of women, insofar as they are institutionalised. Women's political response and agency (their ability to act) is capable of altering the precise form of each structure, allowing for social change, but the structures are beyond the resistance of any individual woman, and shape her experience and condition from birth onwards. The six structures Walby identifies are paid employment (wage labour); household production (domestic labour); culture (including the mass media); sexuality; violence; and the state (through political structures, legislation and the implementation of law and policy). As later chapters show, in our view all these dimensions must be examined if we are to reach a full understanding of women's place and places in contemporary cities.

Post-modernist Feminism

Post-modernism has introduced important new thinking into both social theory and cultural analysis. In social theory, it has challenged the usefulness of both conceptual categories and major theoretical accounts: capitalism, patriarchy, race, gender. The clarity and simplicity of much Marxist and feminist theory is rejected, while social reality is presented as fragmented, differentiated and incapable of simple categorisation. Differences between women (ethnicity, sexual orientation, economic power, cultural milieu) render the concept of 'women' unhelpful, while massive changes in technology, in communication and in access to and control over knowledge have introduced rapid social change on a scale not witnessed since the industrial revolution, with consequent upheavals in communities, families and personal life. Most of the authors of this book reject the most radical assertions of post-modernism — the 'end of history', the rejection of class analysis, and the impossibility of investigating 'women's oppression' in the context of the late twentieth century. But within feminist post-modernism there is a recognition that late twentieth-century social and economic developments may indicate that former ways of thinking are in need of change. Feminism has begun to respond to internal challenges (as, for example, from black feminists) and to recognise and seek to explain cultural and religious differences. Other writers have acknowledged the power of contemporary technological change (Haraway, 1994).

FEMINISM IN ACTION

While these different analytical strands have been emerging throughout the history of feminist thought, feminist activity (whether in the form of an explicit feminist politics, or in the context of personal life or other spheres of activity) has also pursued the aim of overcoming women's subordination in varied ways.

The women's movement

In Britain, feminists first adopted clear aims and pursued these in a political movement during the nineteenth century. 'First-wave feminism' as it has become known, centred on *political rights* (women's suffrage), on *the situation of married women* (campaigning for property rights, for freedom from husbands' violence and restraint, and for parental rights), and on *access to areas of male privilege* (notably to education and to the professions). Feminists also campaigned for improvements in women's working conditions and for women's health needs and their needs as mothers to be met. In this wave of activity, feminists marked up some notable achievements, securing the vote for many adult women in 1918 and for all in 1928, progressively gaining the right to enter universities during the late nineteenth and early twentieth centuries, and gaining important property rights and rights in family law. Through organisations such as the National Union of Women's Suffrage Societies, the Women's Social and Political Union, the Women's Trade Union League, the Women's Co-operative Guild and the Women's Protective and Provident League[5] — women were uniting, identifying political objectives and strategies for achieving them, and working together to secure freedom from oppression and a more equal place in society. Although they were assisted in their campaigns by significant male allies (legislative change before the achievement of women's suffrage — as still today in a male-dominated House of Commons — required male support), most of this campaigning was done by women for women.

The achievement of some of these key objectives placed active political campaigning by feminists on the back burner during the decades which followed — decades dominated by the Depression, by another world war in 1939–45 and by an ideological emphasis on domestic harmony and on complementarity of roles for men and women during the period of post-war reconstruction and emerging prosperity.[6] While feminist activity continued in lower profile activities during this time, as shown below, it was not until the late 1960s, in a burst of academic feminist writing and as women joined in rebellion against the straightjacket of housewifery and deference to continued male power and authority, that another wave of direct political activity emerged.[7]

In second wave feminism the campaign, the strategy and the objectives were different: there was growing disillusion with the strategy dictated by liberal feminism — the pursuit of equality of opportunity through legislative and policy change. Women's suffrage had not delivered equality in society, nor had it brought issues affecting women's welfare to the fore. Access to education and the professions had not delivered equality of access to powerful positions in employment or government, and women's bodies were increasingly objectified in the mass media explosion of the post-war era. The situation of married women was better in terms of their legal position, but marriage still involved exposure to male control and violence for too many women, and the financial dependence most women experienced was disempowering and demeaning. 'Y B A Wife?' asked one of the campaigns of the 1970s, when feminist campaigning embraced campaigns for improved childcare, equal pay in employment, wages for housework, against beauty contests, for access to contraception and abortion, and for an end to sexual discrimination in the provision of goods and services as well as in access to jobs and education. Some lessons from the first campaigns were noted. Women sought not just to create a group of feminist activists, but to develop a broad-based movement of women from all classes through 'consciousness raising' groups and activities. The slogan 'the personal is political' sought to break through the isolation and depression women — and especially housewives — were experiencing, claiming that what happened in the home between women and their families was not understandable in terms of individual psychology or family dysfunction but was a symptom of a wider oppression of women. Women needed to form networks with other women, and to share personal experiences, before they could recognise and act upon their political situation.

Personal life and community activism

The consciousness-raising groups which many women joined in the 1960s and 1970s, often inspired by the Women's Liberation Conferences of 1970–78, and by the new publications for which feminists were responsible (the Virago publishing house, *Spare Rib* magazine and academic journals such as *Feminist Review*), formed a base from which women could draw strength and inspiration in the political campaigning of the years from 1970 onwards. Building from the principle that 'the personal is political', and picking up some issues which had drawn the attention of earlier feminists too, campaigners struggled to achieve equal pay, an end to sex discrimination (including discrimination against lesbians), state support for childcare, freedom and respite from violent homes and from men's violence more generally, control over their own bodies, as well as a peaceful world — in particular one free from nuclear weapons. Women learned that 'sisterhood is powerful', and brought this knowledge into their involvement with trade unions and political parties, forming separate caucuses and campaigning bases where women would set the agenda and run the campaign.[8]

Most importantly, the consciousness-raising movement increased women's confidence to take on issues which affected their own lives.[9] In the cities, women identified the needs of their own communities, and set up Black Women's Centres, campaigned for better housing and tenants' interests, pushed for resources for community centres with nurseries and disabled access, and joined together to support activism for safe transport and to 'reclaim the night'. Rape Crisis helplines were set up, and women worked together to secure premises for women's refuges, where women and their children could be safe from domestic violence.[10] They lobbied local authorities for finance for some of these activities, and achieved some success into the 1980s, especially in authorities where there were Women's Committees or Equal

Opportunity Committees (Goss, 1984; Goss, Stewart and Wolmar, 1989).

Women have also made major changes in their personal lives. Many women sought escape from miserable marriages, and lesbian choices could be more open in some cities, as cultural diversity became a more prominent and valued aspect of city life. To be different could still be dangerous, but more women felt sufficiently confident of their economic status and personal autonomy to make such choices. The city provided opportunities for lesbian social and cultural activities to develop, and for supportive networks to emerge. Most women continued their involvement with men — as family members and as sexual partners — but the terms of the relationship had subtly altered. Women's expectations about employment, education and domestic work, and men's stated views about these matters have all altered significantly in the past twenty-five years, in the direction of greater equality, and of men's increased involvement in domestic life, especially in relation to childcare and parenting. Against this must be set continuing evidence that women perform most domestic duties, although they are more likely than were their mothers to get help with shopping, cooking and washing up. But women with partners increasingly bear a dual burden of responsibility for domestic tasks and income, and the growing numbers of female single parents often have no one with whom to share these tasks.

Our knowledge of women's lives has increased greatly in the post-war years, and feminists have been responsible for drawing the attention of politicians, social scientists and cultural commentators to what Simone de Beauvoir (1972) had called *the second sex*: indeed, many feminists joined the ranks of these groups themselves, and effected change from within. The 'evidence' about women's lives has itself been the subject of important debate and some controversy, however, and it is to this subject that the chapter lastly turns.

FEMINIST 'EVIDENCE': QUESTIONS OF FEMINIST RESEARCH AND METHODS

This book is a product of feminist endeavour, and draws upon the work of its authors and of a much wider academic base of feminist scholarship and contribution. 'Feminist research' is now taught as a subject in many social science departments, and there is a growing literature about feminist research methods and the theoretical ideas from which they spring.[11] The debate about whether there is a form of research into women's lives which can only be conducted by women and which is distinct from all other forms of research is one which must be left to others.[12] But the characteristics of feminist research can be summarised, and are widely agreed.

Feminist research takes note of the relationship between the researcher and those being researched, and seeks to be aware of and to minimise power relations and the potential for exploitation. Thus, quite contrary to the teachings of positivist quantitative researchers, the keeping of 'distance' between interviewer and interviewee, and the use of standardised questions and techniques is to be avoided. Stanley (1990) has given emphasis to the idea that feminist research produces 'unalienated knowledge' — 'that which concretely and analytically locates the product of the academic feminist labour process within a concrete analysis of the process of production itself' (1990, p. 12). Feminists like Finch (1993) and Oakley (1981) have also emphasised the relevance of personal experience in producing feminist knowledge. The researcher need not — indeed should not — pretend to be an objective machine in the collection and analysis of data, but must recognise that the subject of enquiry, the conduct of research and the interpretation of results are all matters affected by personal experience and life events. Oakley's influential works on motherhood, for example, can best be understood, and could only have been produced, in the context of Oakley's own experience of that state.

Feminism also tends to lead the researcher towards 'action research' — where the point of the work is not to map what has happened, or to detail it 'just for the record', but is rather better to understand, and hence to have an influence upon, actual lived reality. In community work, and to some extent in policy evaluation, action research has become an accepted way of working. The researcher does not seek to examine a social phenomenon by 'preserving it in aspic', but to understand and to influence processes of social change, and for feminists, of course, this influence will be in the direction of eroding women's subordination.

Feminist research can perhaps be best summarised in the three words 'seeing, knowing, being'. Stanley's work elaborates these aspects: feminism is more than just a 'perspective' on the world; it is more than simply a way of 'knowing about the world'; in the final analysis it is also 'a way of being in the world' (1990, p. 14).

This book, *Changing Places: Women's lives in the City*, which we now invite you to explore is one in which you will discover a feminist perspective on women's lives in contemporary cities; it is one in which we have assembled a considerable array of knowledge about women's place and women's places in such cities; and it is one which we hope will sustain and influence women's 'ways of being' in those cities.

12

In Chapter 2, Sue Yeandle explores women's work in urban settings, noting the types of work women perform and the conditions under which they work. Carol Walker looks in Chapter 3 at women's economic situation, focusing on those women who experience economic disadvantage and poverty. In Chapter 4 (Section 2), Jane Darke asks us to consider how women are housed and find living spaces, while in Chapter 5 she explores the meaning of home for women. Penny Lidstone's study of what happens to homeless women seeking local authority housing in Chapter 6 concludes this section, raising issues of critical contemporary importance. In Section 3 questions of transport and safety are examined by Rosalie Hill (Chapter 9) and Helen Morrell (Chapter 8), following Jane Darke's contribution (Chapter 7) which looks more broadly at how cities work for women, noting that they can be perceived both as places where danger lurks for women, and as places where individuality and liberation can be achieved. In Chapter 10 Dory Reeves considers how women shop, reviewing a wide range of developments in retailing and in women's shopping behaviour, while Eileen Green, in Chapter 11, develops insights from her work on women and leisure, asking both what leisure is for women, and in what circumstances it can be a meaningful concept for them. Chris Booth's concluding Section 4 suggests there is scope for 'changing places'. The ambiguity of the phrase — referring both to the possibility of changing the urban environment in which women live and to the potential for women to play a more prominent role in the professions which serve and shape urban communities — is deliberate.

NOTES

1 In this chapter, for ease of expression, the world 'subordination' will be used to indicate this 'disadvantage, subordination or oppression', although it is recognised that these terms can be used to indicate specific political and theoretical positions within feminism.
2 For an excellent account of Wollstonecraft's life and work see Tomalin, C. (1992).
3 See, for example, the work of Beechey (1977, 1978) and of Bruegel (1979).
4 This is developed in the work of Benston (1969) and of Dalla Costa and James (1972), the latter developing their theoretical argument into a campaign for 'Wages for Housework'. Domestic work is also explored in some detail in Barrett (1980).
5 For accounts of the suffrage campaign, including sources for the organisation of feminist activists, see the bibliography in Liddington and Norris 1978: 292-296. Boston (1980) and Drake (1984) describe the organised women's labour movement.
6 Finch and Summerfield (1991) provide a good account of marriage and family life in the post-war period.
7 This period is the subject of Coote and Campbell (1982)'s account of the

struggle for women's liberation.

[8] Coote and Campbell (1982) provide an overview of the women's liberation movement in the 1960s and 1970s.

[9] Some of these activities are discussed in Holland (ed.) 1984.

[10] For a detailed account of the women's aid movement in the UK and the USA, see Dobash and Dobash (1992).

[11] Examples include: Roberts (1981); Stanley (1990); Nielsen (1990).

[12] See, for example, the debate in *Sociology*, May 1992, Vol. 26, No. 2.

SECTION 1
Women, Work and Income

2.

WOMEN AND WORK

Susan Yeandle

INTRODUCTION

In contemporary cities, much of the work done by women is visible to the most casual observer: women work in shops, cafés, banks, schools, offices and hospitals, while all city dwellers encounter women workers in going about their daily lives. Some work done by women attracts less notice: home-based workers, for example, are unrecognised at various levels. Most domestic work done in a person's own home is unpaid, and homeworkers, working for money, are often doing work which is unrecorded, unofficial and non-unionised, although both kinds of work are subject to certain kinds of regulation.

This chapter is about the extent and nature of women's work, and about how the conditions of their work have changed and will continue to alter. Labour is a defining characteristic of human societies, and the social organisation, control and definition of labour are critical elements in understanding the status, position and power of those who perform it. This chapter reviews the historical background to conditions for women in the contemporary labour market in Britain and similar industrialised societies. It summarises the current position occupied by women workers, paying particular attention to the type, level and rewards of the jobs they hold as employees. It concludes by considering what the future will look like for women workers in cities, drawing on contemporary debates and theories about changes in the occupational make-up of modernised societies and in the shifting boundaries of class, gender and ethnicity which will dominate the social and urban context in the early twenty-first century.

THE HISTORICAL BACKGROUND: WOMEN'S LABOUR IN THE PAST

Reference to the history of women's work in Britain, or indeed in Western Europe more generally, reveals that from about the sixteenth century onwards, cities and their development have been dynamic

16

factors in the shaping of women's labour. In pre-industrial times women had relatively clearly defined roles in agriculture, cottage industries and family-based production. These tasks were integral parts of a land-based rural economy and, for women of peasant or small land-holding classes, meant involvement in the tasks of dairying, poultry and animal husbandry, kitchen gardening, some aspects of harvesting, simple textile and clothes manufacture, cooking and household tasks. The tasks varied according to locality, terrain and social status, and women in more affluent circumstances progressively moved away from productive tasks and into household management, leisure activities, 'feminine accomplishments' and cultural activities.

From the sixteenth century cities have been centres for the marketing and manufacture of goods, for trade and commerce and, from the late eighteenth century especially, of rapid population growth, linked with the development of industrial capitalism. In the expanding city populations, women found work and 'positions' in trades such as laundering, millinery, garment manufacture, food preserving and production, retail and, above all, domestic service.

In craft and merchant families in the sixteenth to eighteenth centuries, women had fulfilled ancillary roles and had sometimes worked closely with their husbands in the training of apprentices, yet their work had been mainly within the household, and subject to familial control — usually in the form of the more or less benign patriarchal influence of husbands and fathers. As factories and other centres of industrialised activity multiplied in the late eighteenth and nineteenth centuries, specialised roles and tasks emerged for women workers, often away from the supervision of fathers and husbands, in which women worked mainly alongside other women, although usually for male employers and often supervised by men.

The industrialising city offered women some opportunities and freedoms which their rural mothers and grandmothers had not experienced. It also subjected them to new dangers, forms of control and exploitation, and it left many women isolated from the support and influence of kin for most of their youth. Domestic service was usually a residential position, with onerous and often physically demanding duties performed throughout the year, over extremely long working days and without holidays. In most cases, payment was made either annually or on leaving one's position (perhaps after a decade), and as Hufton (1995, pp. 71–86) has shown, its purpose was to provide young women with the wherewithal to subsist and subsequently to marry: whether this took the form of the 'bottom drawer', a 'nest egg' or the more formal dowry varied between European countries.

Some women found other forms of employment in their districts: in textile, pin and nail and pottery manufacture, for example, in towns which grew up around factories and industrial centres; while in the larger cities, laundering, garment-making, millinery, cleaning, shop-keeping, the assembly and manufacture of small and light goods (such as umbrellas, boxes, matches, leathergoods, for example), and — for the most desperate — bar-tending, 'entertaining' and finally prostitution (Alexander, 1976; Rendall, 1990) were common occupations.

By the mid to late nineteenth century women workers in the major industrialised sectors had become closely associated with specific tasks in production processes. The more specialised division of labour which emerged as production became more mechanised and sophisticated led to a sharpening of divisions between men's and women's work, while employers and organised workers, especially for example in the engineering trades, operated with notions of 'appropriate' tasks for women, and sought to control who could gain access to skilled, prestigious and better-paid work (Walby, 1986; Bradley, 1989). Gradually, married women were deterred from 'outside' employment, especially in families containing skilled male workers — and for a period between about 1880 and 1960 'outside' employment for working-class women was in many cities associated with poverty and less respectability. Men who could do so preferred to 'support' their wives and to earn a 'family wage', effectively confining them to an exclusively domestic role, although in some districts, especially those associated with textile production, married women continued to make important contributions to household income from their wages. Social policy both reflected and influenced the importance of male breadwinning, as Land (1976) has demonstrated, and as is summarised in her quotation from the *Beveridge Report*: 'the attitude of the housewife to gainful employment outside the home is not and *should* not be the same as that of the single woman. She has other duties.' (Beveridge, 1942, pp. 51 cited in Land, 1976, p. 111.)

The debates about the precise historical development of trends in women's labour or the significance of these developments for the proposal of theories of gender relations need not detain us here as they are covered in detail in other texts (such as Bradley, 1989; Walby, 1986; and Pinchbeck, 1981, for example). However, the points of agreement between historical and sociological commentators need briefly to be noted. In industrialising societies, women's labour was significantly affected by the progressive — though never complete — separation of home and work; the institutionalisation of individual wage labour was a key development affecting the relative economic power of men and women; the emergence of organised labour in the form of trade unions led to improvements in working conditions for employees, but also emphasised divisions between workers, notably along the axes of skill and sex; and state regulation of employment, often viewed as the 'protection' of female and child labour, was a two-edged sword, limiting women's hours and protecting them from certain kinds of exploitation, but at the cost of confining them to more poorly paid and lower status occupations, and frequently placing them in positions of enforced dependency on a husband or father.

The late nineteenth century saw the intensification of previously latent feminist pressures into campaigns for suffrage, access to education and the professions and rights to property and within marriage, which extended into the twentieth century. These, sharpened by the enforced changes in the demand for female labour brought by the First World War, led to some gains for campaigners, and produced some opportunities for professional activity for middle-class women. Concepts of equality of opportunity became more widely articulated and spread to include ideas about equal pay and employment rights. R.H. Tawney, whose *Equality* was published in 1931, was an influential advocate of equality of educational opportunity throughout the 1920s (Sanderson, 1987, pp. 26–8), and although a Royal Commission on Equal Pay was not set up and able to report its findings until 1946, there had been campaigning for equal pay within parts of the labour movement throughout the previous half-century (Boston, 1980; Bradley, 1989). Furthermore, urban environments in the first half of the twentieth century included many established communities, where families lived in proximity to each other, and where, for women especially, contact and assistance between generations gave important supports, especially to young mothers (Young and Willmott, 1957; Rosser and Harris, 1965; Bott, 1957) in the performance of domestic work.

The Second World War, and the changes in technology and communications which emerged from and followed it, crystallised the changed place of women in city work and life. Throughout the mid and late twentieth century the accelerating pace of occupational change and the technological, social and educational developments which underpinned them were to offer a range of more visible and varied opportunities for women in the city, and it is to these that we now turn.

WOMEN WORKERS IN THE POST-WAR LABOUR MARKET: 1945–1975

During the Second World War, women came to public attention as able to work in skilled and physically demanding occupations: they had done so in large numbers and very visibly in most British cities (Summerfield, 1984; Braybon and Summerfield, 1987; Yeandle, 1993). Women workers remained important members of the labour force in Britain and many other Western European cities throughout the post-war decades. However, there was an important social and political emphasis on women's domesticity and family responsibilities during the immediate post-war years (Finch and Summerfield, 1991; Lewis, 1992, pp. 16–26). This echoed pre-war social attitudes and employment practices, for in Britain[1] and some other countries,[2] many women had been barred from employment following marriage, especially in the interwar years.

The declining age of first marriage after 1945 and during the 1950s, and the continuing expectation that marriage would involve quitting paid employment, meant that during the late 1940s and for the following twenty years, many women had a relatively short experience of full-time employment on leaving school (usually between 5 and 10 years), and then became housewives and mothers for 10 to 20 years[3] before gradually resuming employment, often on a part-time basis, and frequently in a different occupation from the one in which they had originally worked.[4] This pattern of working, often called the two-stage or *bimodal* pattern of female employment, became prevalent in the UK, and dominated official thinking about women's work, during the period between about 1945 and 1975. It was the experience in their earlier working lives of many of the older women living in British cities today. Gradually, and increasingly in the following years, women's patterns of working became more intermittent and flexible, reflecting both changes in domestic practice, changes in educational experience and attainment, and changes in opportunities for paid work in the cities (see Tables 2.1 to 2.5 for an overview of women's

Table 2.1 Women's earnings relative to men's: UK 1995

	Females			Males		
	manual	non-manual	all	manual	non-manual	all
Average gross weekly earnings						
(£s)	188.1	288.1	269.8	291.3	443.3	374.6
of which						
overtime payments	12.4	5.2	6.5	43.9	12.3	26.6
PBR*, etc. payments	8.4	5.7	6.2	13.6	17.4	15.7
shift, etc./premium payments	5.3	2.4	2.9	9.8	2.7	5.9
Average gross hourly earnings						
(p)						
incl. overtime pay and overtime hours	464	776	715	644	1133	891
excl. overtime pay and overtime hours	455	775	714	625	1136	897

* payment by results   Source: *New Earnings Survey 1995, CSO, 1995b*

Table 2.2 Women's and men's pay by selected age groups: average hourly pay excluding overtime UK (pence)

Age group	Full-time women		Part-time women		Full-time men	
	manual	non-manual	manual*	non-manual**	manual	non-manual
25–29	482	745	74.5%	82.9%	597	892
30–39	483	875	76.0%	77.3%	652	1161
40–49	461	835	76.7%	80.3%	672	1306
50–59	452	781	77.8%	83.8%	633	1287

* percentage earning less than 460p per hour
** percentage earning less than 800p per hour
Source: New Earnings Survey 1995, CSO, 1995b

Table 2.3 Economic activity rates of women and men, GB 1901–1991

	1901	1911	1921	1931	1941	1951	1961	1971	1981	1991
All women	32	32	32	34	—	35	38	44	48	53
All men	84	84	87	91	—	88	86	81	77	74

Source: Hakim, 1993

Table 2.4 Socio-economic group* by sex (all persons aged 16 and over), GB 1975–92 (%)

	1975	1979	1983	1989	1992
Women					
Professional	1	1	1	1	2
Employers and Managers	4	5	6	9	9
Intermediate & junior non-manual	46	45	46	47	50
Skilled manual & own account non-professional	9	9	9	9	8
Semi-skilled manual and personal service	31	30	30	25	21
Unskilled manual	9	10	9	8	10
Base = 100%	11799	11102	9754	9600	9319
Men					
Professional	5	6	5	7	8
Employers and Managers	15	15	17	20	20
Intermediate & junior non-manual	17	17	15	16	17
Skilled manual & own account non-professional	41	40	39	38	37
Semi-skilled manual and personal service	17	17	18	15	14
Unskilled manual	5	5	5	5	5
Base = 100%	10902	10280	8886	8815	8504

* The socio-economic group shown is based on the informant's own job (or last job if not in employment). Excluding those in the Armed Forces and any who have never worked.
Source: General Household Survey 1992, HMSO, 1994

Table 2.5   UK women: wage earners and salaried employees by activities 1980–1992 (%)

Industry	1980	1984	1988	1992
Agriculture, hunting, forestry & fishing	1.0	0.9	0.8	0.7
Mining and quarrying	0.2	0.2	0.2	0.1
Manufacturing	21.1	16.9	15.1	12.5
Electricity, gas and water	0.7	0.7	0.6	0.6
Construction	1.1	1.3	1.3	1.3
Wholesale and retail trade, and restaurants and hotels	23.9	24.2	23.5	23.8
Transport, storage and communications	3.0	2.9	2.9	3.0
Financing, insurance, real estate and business services	8.2	10.1	12.1	12.5
Community, social and personal services	40.7	42.8	43.7	45.5

Source: OECD, 1994

economic activity, their industrial and occupational distribution, and their average pay in the UK).

Over the whole post-war period, important and far-reaching changes were occurring in the labour market, both in the type and structure of occupations, and in the kinds of jobs and skills which the labour market required and could sustain. The historical divisions between women's and men's work, noted above, had become entrenched into a system of occupational segregation by sex throughout the twentieth century, so that, by 1945, the roles of nurse, primary school teacher, typist, secretary, cleaner and shop assistant, for example, had become almost synonymous with women's work. This *horizontal* occupational segregation by sex was now analysed as existing alongside *vertical* occupational segregation by sex in which men tended to occupy a disproportionate share of the higher skilled, high status and better-paid jobs in a given sector (Hakim, 1981). Hakim's analysis of trends between 1901 and 1971 showed that an overall picture of stability in the degree of occupational segregation by sex disguised the fact that throughout much of the twentieth century slight decreases in horizontal segregation had been offset by small increases in vertical segregation. Thus women were not moving towards a fairer share of the best jobs, although they were slightly more likely to be working alongside men (Hakim, 1978).

In the wake of the Second World War, the most important changes for women workers whose job opportunities were still shaped by this occupational segregation, were the development of the welfare state, with its opportunities for employment in teaching, nursing and social services, and their associated occupations and bureaucracies; the development and (in some cases) subsequent decline of certain types of manufacturing industry (such as the paper and board industry and the electrical assembly and emerging electronics industries, for example); and the massive growth of employment opportunities both in clerical and junior managerial work, and in retail. The introduction of part-time working in some industries during the Second World War gave employers a model for development in the subsequent decades (Yeandle, 1982; Walby, 1986), maximising their use of plant and machinery, capitalising on the relative cheapness of such labour, and calling for labour which women whose primary role was domestic could supply, in twilight shifts and part-time work during school hours. As discussed below, from about the 1970s onwards, part-time work was also introduced into the massively expanding retail sector.

During the middle of the twentieth century the expanded educational opportunities which had been made available to working-class children and to girls in the preceding decades began to affect women's occupational choices and possibilities. The debate about equality of educational opportunity which emerged in Britain in the 1930s and culminated in the 1944 Education Act, making secondary education available to, and compulsory for, all children, was not focused on rectifying girls' disadvantages, but laid the foundation for the discussion in the 1960s and 1970s about sexual discrimination and the right to equal opportunities in education and employment for girls and women. Combined with pressure from the renewed campaigning of feminist activists, this shift led in Britain (with parallels over a similar period in many other European countries, in some cases under the influence of the Treaty of Rome 1957 and the legislation and directives of the European Economic Community) to legislation in the form of an Equal Pay Act 1970, a Sex Discrimination Act 1975 and an Employment Protection Act 1975. Officially, women gained an entitlement to equal access to education, training and employment, to equal pay for equal work, and to maternity rights designed to enable them to keep their jobs during pregnancy and early motherhood if they chose.[5] However, in contrast to some other European countries, parental leaves and rights have in Britain been statutorily available only to women, an aspect of public policy and employment practice which has served to emphasise sexual divisions between workers despite its origin as an equal opportunities measure.[6] Although these Acts have required subsequent amendment and have not delivered equal access to well-paid, secure and rewarding jobs with men,[7] the period from 1970–95 has seen a marked improvement in women's educational qualifications and notably in their achievements in higher education (albeit mainly in subjects and disciplines with which women have traditionally been associated — arts, languages, social science, and disciplines linked with the teaching and health professions). For an important group of women, as will be seen below (p. 26), these achievements have opened up access to professional and managerial jobs, especially in the service sectors of the economy.

The labour market in which women were active in the half-century between 1945 and 1995 was also a sex-segregated labour market, as mentioned above, and besides noting the changes in women's participation, we need also to be aware of the types of work and occupation in which female city-dwellers have been engaged. As this occupational segregation has been a persistent trend, the entire period 1945–95 is discussed here, and will serve to introduce those aspects of the latter part of the period which are considered in the final sections of the chapter.

Between 1945 and 1975, the main occupations in which women were employed included the 'semi-skilled' manual trades (where their share of employment rose from 38 per cent in 1951 to 46 per cent in 1971), sales and shop work (where they were over half the workforce throughout the period), clerical work (where their share of the jobs increased from 60 per cent in 1951 to 73 per cent in 1971) and what official statisticians designated as 'lower professional and technician' jobs, where they held about half of the available positions throughout the period. Within this broad categorisation, women were dominant specifically in nursing and teaching, in typing and secretarial roles, in catering, cleaning and hairdressing, and in product packaging and repetitive assembly work. By 1975, about half of all women with dependent children were in employment, although about twice as many worked part-time as full-time (OPCS, 1980). For most of these women, part-time hours meant 16–24 hours per week, although part-time employment covered a wide range of arrangements, from those working less than 8 hours per week, to those working 30 hours per week (Department of Employment, 1975).[8]

The industries in which women held these jobs varied, but were the ones either traditionally associated with women's work, or were part of the expanding service sector: clothing and footwear, professional and scientific services, distributive trades, insurance, banking, finance and business services, leather and leather goods, textiles, food, drink and tobacco, and instrument engineering. The industries where women were exceptionally few in number by 1975 — mining and quarrying, shipbuilding, construction, metal and vehicle manufacture — were all industries which would suffer further decline and restructuring in the following decades, and it needs to be understood that this uneven industrial distribution (often referred to as the industrial and occupational sex segregation of the workforce) has offered women some protection from the labour market effects of the recessions of the last quarter of the twentieth century.

Domestic work for women in the period between 1945 and 1975 is discussed later in the book in further detail. It is necessary here only to note that the domestic labour of women was affected by several important trends[9] — rising standards of living for most households throughout the period (whether measured by disposable income or quality of housing); increased availability and ownership of domestic appliances (which altered the nature of domestic work, especially laundry, cleaning and food preparation); smaller household and family size and a greater likelihood for poorer women of living in a one-family household; and changes in the allocation of time, as married women and those with children increased their participation in paid employment. The amount of domestic work done did not necessarily decline as households obtained washing machines and vacuum cleaners and other appliances, and as Oakley (1974) showed, most housewives experienced their work as monotonous, repetitive and isolating. The fact that housework was frequently combined with paid work outside the home, or with homeworking,[10] tended to make its performance even more onerous, and throughout the post-war decades, housework and other forms of domestic work remained as tasks which women performed, with the help of their husbands if they were lucky, and under their husbands' control if they were not. Chapter 5 in this volume, which explores the meaning of home for women, draws attention to the way homeworking of this kind affects a woman's experience of her domestic environment.

INDUSTRIAL RESTRUCTURING AND CHANGE: 1975 AND BEYOND

All European countries have witnessed important developments in their occupational and industrial structure in the last quarter of the twentieth century. In Britain, the decline of traditional industries such as coal, steel, shipbuilding and heavy engineering had begun long before 1975, but during the 1980s it became rapid and apparently irreversible. The main jobs lost were men's, although the effect on women of these changes has been equally important. Most of those displaced from such industries were married men, often with craft skills, whose expectation was of lifelong employment in the trade to which they had been apprenticed in youth. The effect of their redundancy and in many cases prolonged subsequent unemployment on their wives' labour (both paid and unpaid) has been significant (Morris, 1987, 1993). Men's presence in the home and reduced household income both have important effects on women's domestic work. Although some greater sharing of domestic chores has been recorded in some cases, the structure of social security benefits has led many low-paid women to resign their jobs while their husbands are out of work, and in many situations women's work has become more onerous, with the increased pressure of managing an extremely tight family budget. Carol Walker discusses these aspects in more detail in the following chapter.

Other related changes have also affected women's paid employment. Although in 1980 over one-fifth of all female employees were working in manufacturing industry, for example, that figure had declined to one in eight by 1992, while higher proportions of women were working in both the business services sector (12.5 per cent in 1992 compared with 8.2 per cent in 1980) and in the social and personal services sector (45.5 per cent in 1992 compared with 40.7 per cent in 1980: (OECD, 1994). Thus alongside the decline of traditional industries, there have been important developments affecting, especially in Britain, work in sectors which have proved attractive to women employees. Many of the new jobs have been in the service sector and in office work. These jobs include many in which levels of pay are low and opportunities for career progression very limited, although most of the jobs are dependent upon a basic level of education, such as is achieved by the majority of school leavers. Many of these jobs are part-time, and are held by women who are also performing significant unpaid labour in the home. Recent calculations show that in the UK 41 per cent of women workers were employed for fewer that 30 hours per week, 31 per cent for between 37 and 42 hours per week, and 9 per cent for 45 hours per week or more (European Commission, 1994).

Over 5 million UK women work part-time (*Social Trends*, 1995, Table 4.12), and many, whether full- or part-time, have contracts which involve flexible working patterns. About 1 in 5 of employed women, regardless of their weekly hours, has a contract permitting either flexible working hours or term-time working, while 6.4 per cent of full-timers and 5.3 per cent of part-timers are employed on an annualised working hours basis (*Social Trends*, 1995, Table 4.16). Part-time employment is especially important among women with school age children, and it is the most important economic activity status for mothers of children aged 5–15 years (*Social Trends*, 1995, Table 4.5). By 1995, for example, most of those employed by the supermarkets, chain stores and retail parks which have developed in the last thirty years are women workers, the majority of whom work part-time (Penn and Wirth, 1993), often at weekends and in the evenings, and frequently also responsible for bringing up children and for family organisation and domestic work. As subsequent chapters show in more detail, these developments in retailing have come to dominate city centres and to influence major aspects of city life, including transport systems and the nature of local communities.

By contrast, there are expanding opportunities in the service sector for employment in managerial and professional roles, and women have secured a slowly increasing proportion of these positions in recent years. By 1994, just over 1 in 10 UK employed women were in jobs designated as 'managers or administrators', with a further 1 in 12 in 'professional' jobs. Additionally, 1 in 10 were in 'associate professional and technical' occupations (OPCS, 1995). These are jobs requiring education, qualifications, skills and expertise, and pay women wages on which it is possible to support themselves (although not necessarily their dependants). Men still occupy far more of these positions than women (Arber and Ginn, 1995), but women in such jobs have better access to maternity rights (McRae, 1991), pensions, and other occupational benefits, as well as more security of employment and opportunities for career advancement, than do women in lower-paid service sector jobs.

Women in the latter jobs usually combine their employment with unpaid domestic work within the family — whereas increasingly, women in better paid occupations, who are often in dual-earner households, employ other women to do some of their household's domestic work (Gregson and Lowe, 1994). The trends here are opening up an increasing divide between women's material conditions and life chances: cities are increasingly populated by a polarising female workforce. At one pole women have demanding careers, good salaries and benefits, often accompanied by their own cars, smart wardrobes, and access to modern household conveniences and good quality prepared food. Their lifestyle is supported by paid service workers who clean their homes, mind their children, do their laundry and look after their gardens. At the other end of the scale, women hold multiple low-paid jobs, including paid domestic labour in others' homes, manage complex domestic routines[11] shaped by their children's school hours and in some cases by their husbands' working days, and frequently work unsocial hours in the evenings and at weekends. Women in the latter group have intermittently to drop out of employment when family demands change, when male partners fall out of work or leave, or when work opportunities collapse. They are on the dividing line between dependency on state benefits and self-maintenance through paid employment, and the few resources available to change their position (through access to education as mature students, for example) are limited and inadequate.

Such divisions have been further reinforced by the emergence during this period of unemployment as a phenomenon affecting married women and mothers as well as single women (Coyle, 1984; Martin and Wallace, 1984). This is not to suggest that women did not experience unemployment in the past, for there is evidence that they did (Factory Inspectorate, 1878–1974; Alexander, 1976; Walby, 1986), although the casual status of many women's employment often made their unemployment or underemployment difficult to record. Indeed, it has long been recognised that much of women's unemployment is 'hidden' or at least masked by the other activities which women undertake, and especially by their role in household and caring tasks. Official statistics of claimant unemployment show that women in this category were few throughout the early 1970s (around 100,000), but that after 1975 there was a gradual increase, reaching a peak in 1986 of approximately one million women, and standing at well over half-a-million in the mid-1990s. The broader context for these developments is the focus of further discussion in Chapter Three.

In recent decades, this increased male and female unemployment has produced a sharpening divide between households with two earners, living in owner-occupied housing and sufficiently well-off to take annual holidays, own cars and participate in sports, leisure and entertainment activities; and households affected by long-term unemployment, where all income may derive from welfare benefits, where housing is likely to be rented, often in the public or quasi public sector, and where mobility and activities are severely hampered by lack of funds. Women's relatively poor access to occupational training,[12] and the poorer occupational and educational qualifications of women over 35, confine some of them to low-paid and low-status sectors of the labour market (as is shown in the section which follows) interspersed with periods of unemployment which may extend in some cases for years. These women, however, are not idle: their work is mainly in and for the household, in making ends meet and in coping on their low incomes, in shopping where items are cheapest rather than most convenient, and in taking time to carry out these tasks. They also have more extensive responsibilities for the daily care of other kin than do more affluent women (Land, 1991). On public transport, the ferrying of children to school, the shopping tasks for the week, and the visits to hospitals, doctors, benefit offices, etc. all take up more of the available time than is the case for better-off women who can drive, often in their own cars, as they rush to complete their busy schedules.

Polarisation in the female labour force is further complicated by divisions based in ethnicity and culture which are both the result of prejudice and discrimination and of varying cultural practices. Although cities are where most women from ethnic minorities live, their distribution is highly variable, and this makes generalisations about their work situation problematic. It is important to differentiate between ethnic minority groups,[13] as there are notable variations. Data for 1993 has shown (Sly, 1994a) that ethnic minority women in Great Britain represent 6.1 per cent of the female population aged 16–59 years. The percentage is higher in younger age groups, rising to 7.5 per cent of 16–24 year old women. Although black and Indian women have national economic activity rates similar to those of white women (66 per cent and 61 per cent respectively, compared with 72 per cent for white women), women from the Pakistani/Bangladeshi group are much less likely to be economically active (25 per cent). Economically inactive women in the latter group are much more likely than others to be working unpaid in the home, looking after their families (72 per cent compared with 47 per cent and 56 per cent of economically inactive black women and white women): and it is 'inactive' black women who are the most likely to be students: 25 per cent, compared with 10 per cent of the Pakistani/Bangladeshi group, and 15 per cent of the white group (Sly, 1994a).

Unemployment is not evenly distributed among women either: while 8 per cent of white women were unemployed (using the ILO definition of unemployment), the figure for all ethnic minority women was much higher (13 per cent) and rose to 20 per cent in the case of black women. This difference applies whether or not women hold qualifications: for example, the ILO unemployment rate for white women aged 16 to 59 years was 4 per cent in 1993, compared with 9 per cent for women in ethnic minority groups. This difference suggests that discrimination on the grounds of ethnicity is a factor with which minority women are faced.

While all women in employment are strongly concentrated in the service sector, there is variability here too: 83 per cent of white women, 91 per cent of black women, and 77 per cent of Indian women were employed in this part of the workforce in 1993. However, Indian women are much more likely than any other group to have jobs in the manufacturing sector (21 per cent). Minority women in paid jobs are more often full-time employees than are white women — 40 per cent of women in employment worked part-time in 1993 compared with 25 per cent of women in the ethnic minority population of working age. And although overall, white people of working age are more likely than those from other ethnic groups to hold formal qualifications, the trend in recent years has been towards convergence: in 1989–91, for example, 62 per cent of Pakistani/Bangladeshi women had no qualifications, compared with 72 per cent five years before. Furthermore, 21 per cent of the Indian population of working age held higher qualifications in 1993, compared with 19 per cent for the white population (men and women).

CONCLUSION

Women's work is an increasingly visible and essential part of the economy: women continue to perform most of the unpaid work on which the comfort of the population depends, and play a major and growing part in the most vibrant sectors of the economy. There are growing divisions between women in terms of pay, benefits and occupational status, and important differences between women of different ethnic and cultural origins.

Work and employment play a central part in the lives of women in cities. Just as for rural peasant women in the European past work and life were essentially the same thing, so for women in modern European cities, existence is dependent for most upon labour: labour which takes many different forms, is variably rewarded, and which may be carried out in a range of combinations. Nevertheless, work for women in modern cities involves some experiences which are common if not universal, and which have altered the character of city life. For many women, employment means travelling from housing estates and suburbs into town centres, and paid employment gives a specific focus to their lives. It brings them into the centre of city life as workers, travellers, consumers, and customers. Women workers fill shops, banks and streets during their midday break, when they are often doing domestic chores essential to family living, and this gives a particular character to city spaces at these times. By contrast, fewer women of working age than ever before are available to shop locally for daily requirements, and this has changed the nature of social interaction in residential communities as well as making the housing estates of suburbs and residential areas relatively deserted places, where elderly residents and those caring for small infants are the most visible, for long periods in the working day. These consequences of women's changed employment patterns are discussed in more detail in subsequent chapters.

NOTES

[1] The marriage bar in Britain is discussed by both Walby (1986) and Bradley (1989). Its informal use by small employers has also come to light in small-scale studies (Yeandle, 1987).

[2] For example, Frevert (1989) notes that in Weimar Germany, under the 1932 Law on the Position of Female Civil Servants, 'the civil service was legally bound to dismiss married female employees' (pp. 197–8).

[3] Italian women were also pressed into what Saraceno (1991, p. 190) has described as the 'pattern of the institutionalized life course of women as wives-mothers, which in Italy lasted from the late 1940s to the end of the 1960s'.

[4] Surveys of women's employment which document these trends in some detail include Hunt (1968) and Martin and Roberts (1984). Yeandle (1984) demonstrates the extent to which varied occupational experience became a feature of women's working lives during the 1960s and 1970s.

[5] The main aspects of the legislation were outlined by Hewitt (1975) in a contemporary publication.

[6] Lewis (ed.) (1993) gives a recent overview of employment and family policies in Europe, and contains a particularly useful introductory chapter.

[7] See Rendel (1978) and Snell (1986) for an account of the original legislation; Lewis (1992, pp. 65–91) for a succinct account of women's employment and pay in the post-war period; and Pillinger (1992) for an analysis of women's

pay and employment in the European Community.

[8] During this time, many married women paid reduced National Insurance contributions, thereby foregoing access to employment-related unemployment and sickness benefits. This option was phased out after 1976 following the passage of the Social Security Pensions Act 1975 (see Chapter 7 in Mackie and Pattullo, 1977).

[9] See Crow (1989) for a discussion of how the 'domestic ideal' developed during this period.

[10] Allen and Wolkowitz (1986) review the evidence on the extent and economic importance of homeworking, which began to emerge in the 1970s, and to which their research made an important contribution. Tasks typically done by homeworkers include the assembly of goods (e.g. toys, Christmas crackers, circuit boards), packing, knitting, typing, and sewing.

[11] Recent evidence on the performance and management of household tasks shows that in most homes headed by married couples or a couple living as married, it is still mainly the woman who carries out major domestic duties: 84 per cent for washing and ironing; 70 per cent for making the evening meal; 68 per cent for household cleaning; 60 per cent for looking after sick children. In these examples, most other couples claim to share the tasks equally: it is rare for men to carry the main responsibility. (*Social and Community Planning Research* quoted in Social Trends 1995, Table 2.7.)

[12] Recent figures (from *Employment Department*, quoted in Social Trends, 1995) show women receiving slightly more job related training than men in some age groups, but a decade ago there were marked disparities, especially for young workers, and together with lower levels of educational achievement in the past, and a history of marked sex segregation in occupational training patterns, there is an accumulated disadvantage which women as a category have still to overcome.

[13] The groups differentiated here are those used in the collection of Labour Force Survey data. For a discussion, see Sly (1994a).

3.

THE FEMINISATION OF POVERTY: WOMEN AND SOCIAL WELFARE

Carol Walker

INTRODUCTION

The latest figures available at the time of writing show that in 1992–93, between 13 and 14 million people (about one quarter of the population of the UK) were living in poverty: more than double the number affected in 1979. Poverty is not confined to any particular group or to any particular place and yet it is not a 'random' experience (Glendinning and Millar, 1992, pp. 2, 3); certain groups are much more likely to experience it than others. This chapter considers why women are more vulnerable to poverty than men and the different ways poverty impacts upon their lives. It considers also how living in a particular environment, the city, which has long been closely associated with poverty, can exacerbate their experience of deprivation and social exclusion.

POVERTY AND THE CITY

The relationship between the city and deprivation has been well documented in Britain and other countries for over a century. Many of the first, pioneering studies of poverty in Britain were concerned with urban poverty (Booth, 1903; Rowntree, 1901). In many people's minds poverty and deprivation have become synonymous with the inner city. As MacGregor and Pimlott (1991, p. 15) point out, despite the proud history of many of our cities as creators of wealth: 'It is a bitter indictment of our own time that the phrase "inner city" should today universally conjure up images of disorder, poverty, fear, vandalism and alienation'.

No urban area, city, or even inner city, however, should be regarded as homogeneous. They are places of great diversity, where affluence and poverty live 'cheek by jowl' (Oppenheim, 1993, p. 153) and each urban area has its own ' "internal geography" ... which helps to produce distinctive patterns of poverty and deprivation' (Goodwin, 1995). Thus there are differences both between and within urban areas.

There have been numerous government sponsored initiatives in the UK to try to reverse the decline in the inner city and yet urban areas remain the most deprived parts of Britain while the gap between them and the rest of the country has widened (Oppenheim and Harker, 1996). Gordon and Forrest's (1995) analysis of the 1991 Census shows that 19 out of the 20 most deprived areas are urban areas, 15 of the 20 in London. While Goodwin (1995, p. 71) argues that there is some ambiguity about which cities are the most deprived, he notes '. . . a gradient of deprivation, running downwards and outwards from the cores of Britain's large metropolitan areas. Thus, the inner areas of London, Birmingham, Glasgow, Liverpool and Manchester regularly appear at the top of most tables of urban deprivation and poverty.'

Urban areas have suffered most from the collapse of industry in Britain. Goodwin (1995) draws together 'a broad picture' of the deprivation many face: official unemployment rates of over twice the national average; claimants receiving income support running at almost three times the national average; overcrowding up to four times the national rate; and higher levels of ill-health and mortality, a growing gap between these areas and the rest of the country and widening social and geographical inequality between the richest and poorest boroughs within each city (Townsend, 1987).

Higher than average levels of deprivation are not confined to the inner cities themselves: in many areas they have spread to outer urban estates. These estates, which were planned to provide overspill accommodation from the overcrowded and decaying inner cities in the 1950s and 1960s, were intended to provide a more positive environment. However, in many cases they merely compounded the existing problems imported from the inner city by putting together socially homogeneous, and usually vulnerable, people in an environment which lacked employment opportunities, shops, transport and play facilities for children in 'The grim and monotonous system-built environment of the estates themselves [which] adds to the pervading sense of despair, as does their relative isolation and distance from city centre services and facilities' (Goodwin, 1995, pp. 71–2).

Townsend paints a grim picture of poverty and deprivation experienced by those living in the most deprived parts of the city:

> [they] suffered disproportionately, not only from poor housing, lack of household facilities and possessions and inadequate clothing, but also from risk of road accidents, litter problems and lack of garden and play facilities. They also moved house more often and had more health problems than those residents in less deprived areas . . . . Many feared eviction, were concerned about isolation and experienced racial harassment, and almost one-fifth had encountered street or estate violence in the previous 12 months. Just over a fifth said they suffered from poor

public transport, while just under a fifth considered that they owed money.
(see Goodwin, 1995, p. 77).

While the problems described above are not experienced only by women
they are experienced especially by women. Women have fewer
opportunities to escape their environment either for work or leisure and
their caring responsibilities ensure that they are constantly confronted
in their everyday lives with the lack of opportunities for them and their
children. Living in poverty amidst affluence puts additional pressures
on the domestic manager, who in poor households is usually the
woman.

POOR WOMEN DON'T COUNT

The Department of Environment's 'official' definition of urban
deprivation (Goodwin, 1995) includes a combination of eight indicators:
unemployment, lone pensioners, lone parents, population change,
mortality, ethnic origin, overcrowding and shared amenities. Women are
clearly over-represented in the first three of these, the causes and impact
of which are discussed later. They tend also to be more affected by the
remaining indicators. Poor women experience, like all poor people,
higher morbidity and mortality rates than the more affluent — as do
their children; black women are further disadvantaged by the
discrimination and exclusion they suffer on account of their 'race'; and
as managers of the household and primary carers, women face the full
brunt of poor housing as exemplified by overcrowding and shared
amenities: 'Not having hot water or a bathroom means different things
to the man who uses hot water for washing and shaving in comparison
with the woman who is responsible for childcare, the washing of clothes
and cleaning the house' (Payne, 1991).

The discussion of the feminisation of poverty has received greater
attention in recent years (Scott, 1984; Rein and Erie, 1988; Glendinning
and Millar, 1992), however the poverty of women is not new. As Lewis
and Piachaud point out (1992), at the beginning of the twentieth century,
61 per cent of people on poor relief were women; in 1992 60 per cent of
people on income support (the state means-tested safety net) were
women (Oppenheim, 1993, p. 93). However, the extent of women's
poverty was often neglected in early studies of poverty (Glendinning
and Millar, 1987) and remains neglected in the official statistics.

In the post-war period, governments have published two separate
series of statistics which have been used by commentators, though
increasingly reluctantly, if at all, by government ministers, as measures
of poverty. The first, the *Low Income Statistics* (DHSS, 1988) presented

information on the number of people living below, at or slightly above social assistance levels (named in turn national assistance, supplementary benefit and income support). The Conservative government replaced these with a new series on *Households below average income* (HBAI) (DSS, 1995) which shows, first, the number of individuals in households with incomes below various thresholds and, secondly, the number of individuals living in households in the bottom five deciles of the income distribution and the rises in real income each group experiences. Both these measures underestimate poverty and low income among women because they are based on the assessment unit or household income and not individual income. In one of the most authoritative analyses of the HBAI statistics Goodman and Webb (1995, p. 3) make light of this important shortcoming: 'If... one household member received all the income and denied other household members any benefit from it, then it would clearly be a mistake to measure each individual's living standards on the basis of total household income. In practice, however, this extreme assumption is unlikely to hold.'

While the total refusal to redistribute any resources within the family unit may be 'extreme' (though still not unknown), there is now substantial evidence to show that some members of a household do experience poverty when the distribution of resources — whether by greed or convention — is inequitable. Vogler (1989), for example, found that only 20 per cent of households used an 'egalitarian pooling system' in managing household income and that generally within couple households, women commanded fewer resources (see also Pahl, 1983).

The interaction between the rules of the benefits system and women's experience of unemployment means that women are also under-represented [according to some commentators by as much as a quarter (Callender, 1992)] in the government's official unemployment count which includes only those people registered as unemployed and claiming benefit. For reasons discussed below many women who would like to work and/or are looking for work are ineligible for benefit. In a couple, where both are unemployed only one is counted as unemployed, and for reasons again discussed below that one is likely to be the man. Where the husband or male partner is unemployed, women are often deterred from seeking work because of the punitive earnings rules attached to benefit entitlement. The proportion of non-working women with working partners entering employment rose from 14 per cent to 21 per cent between 1979 and 1993. By contrast the proportion of non-working women with unemployed partners who moved into work actually fell from 15 per cent to 9 per cent (Harker, 1996, p. 12). In addition many working wives give up work when their partner loses his job (Moylan *et al.*, 1984; Morris and Llewellyn, 1991).

The Labour Force Survey count of unemployment shows much higher rates of unemployment among women than the claimant count. For example in autumn 1994, 24 per cent (576,400) of the claimant count were women. By contrast according to the Labour Force Survey definition (i.e., people who did no paid work in the previous week, had actively looked for work in the last four weeks and were free to start work in two weeks, regardless of whether they were claiming benefits), 34 per cent (846,800) were women. If the Labour Force Survey's alternative and less stringent definition of unemployment is used (i.e., people who want work, are available, but have not looked in the last four weeks) then 63 per cent (a further 620,600 women) were unemployed (Oppenheim, 1993, p. 95).

THE EXTENT OF POVERTY AMONG WOMEN

Because the major sources of data on poverty are not broken down by sex, it is only possible to make rough estimates of the extent of poverty among women. Oppenheim and Harker (1996) estimate that in 1992 approximately 5.4 million women were living on incomes on or below income support level compared to 4.2 million men and that, according to this measure, around 56 per cent of adults living in poverty were women. Webb (1993) found that in 1991 two-thirds of adults in poor households were women and that the women in these households had about half as much independent income as men — £99.90 per week versus £199.50) (see Table 3.1). Having children considerably increases a family's vulnerability to poverty. Between 1979 and 1992–3, the risk of poverty increased from 8 per cent to 24 per cent among couples with children and from 19 per cent to 58 per cent among lone parent families (Harker, 1996, p. 8). In 1992–93 over 4.3 million children (33 per cent of all children) were living in households living on less than half average income. According to the same definition couples with children account

Table 3.1 Mean independent income of women, 1991, by source and by family type

	[Non-pensioners]				[Pensioners]		
Income source (£ per week)	Single with children	Married with children	Single no children	Married no children	Single	Married	All women
Earnings	37.00	54.70	92.40	90.60	4.20	6.10	53.40
Self-employment	4.70	5.10	2.70	7.10	0.20	0.40	3.80
Social security	61.50	16.40	12.20	3.50	57.40	30.40	23.40
Investments	2.80	5.00	6.70	13.10	18.90	18.20	11.00
Pensions/annuities	2.60	0.20	2.10	1.50	19.70	6.50	4.90
Other	15.20	3.80	4.10	3.40	0.80	0.60	3.40
Total	123.80	85.10	120.00	119.10	101.20	62.20	99.90

*Figures have been rounded   Source: Oppenheim and Harker (1996), p. 93

for the largest group in poverty (37 per cent in 1992–93) and lone parents now account for the second largest (17 per cent). The overall risk of poverty to lone mothers is extremely high (Oppenheim and Harker, 1996, p. 95).

WHY ARE WOMEN POOR?

'Poverty is a consequence of an inability to generate sufficient resources to meet needs' (Glendinning and Millar, 1992). While women may find themselves unable to do this for a number of reasons — most notably a lack of suitable employment — which also affect men, they are further disadvantaged by a number of factors related to their gender. In particular women's access to, and participation in, the labour market is severely circumscribed by the genderisation of work and by their role as carers. This is discussed in the previous chapter.

Despite their increased participation in the labour market in the post-war period, most women retain the role of primary carer — not only of children but of other family members who may need care, even when they are not the nearest blood relative (Qureshi and Walker, 1989). Because of their caring responsibilities, many are restricted to work which is close to home, with flexible hours and which does not require time away from home. Many women endeavour to combine work and caring responsibilities by working part-time. Such domestic ties can restrict labour market participation of women living with a partner, they are of course even more significant for women bringing up children alone.

Low pay is now one of the main causes of poverty in Britain for men and women. However, women are more likely to be concentrated in low-paid jobs. Despite increased participation by women in the labour market, the workforce has remained remarkably segregated (see Chapter 2). Although equal pay and anti-discrimination legislation has led to a fall in the gap between men's and women's earnings, in 1994 women's full-time gross weekly pay was 72 per cent of men's. Women accounted for two-thirds of those earning less than the Council of Europe's decency threshold (£5.88 an hour) in 1994. Most part-time workers are women and over three-quarters of them were low paid in 1994. Black women are likely to have even lower hourly earnings than white women, though some boost their income by working more hours or shift work (Oppenheim and Harker, 1996, p. 95). Being lower paid or in part-time employment women are likely to have less employment protection, fewer occupational benefits in sickness or after retirement, fewer fringe benefits while in work and to be vulnerable to unemployment.

While the link between unemployment and poverty has been firmly established, it is still often assumed that women's unemployment is less serious and causes less financial hardship than is the case for men (Callender, 1992). As discussed earlier, rates of unemployment among women are significantly higher than the official statistics indicate. Unemployment rates among black women are substantially higher, averaging 16 per cent in 1994 compared to 7 per cent for white women (Oppenheim and Harker, 1996, p. 115). Changes in social security payments for the unemployed, in particular the introduction of the Job Seeker's Allowance which will increase dependence on means-tested benefits, will further disguise the true rate of women's unemployment.

The extent of financial hardship caused by unemployment varies with the economic circumstances of the household and is most devastating where the woman is a sole earner. However, most households live to their income, so the loss of part of that income will almost always have an effect on living standards, particularly in low income households where there is no financial cushion and few savings. The Low Pay Unit have calculated that there would be one million more families living in poverty if both partners were not working (Walker, 1993). Thus the loss of the woman's wage, even where it is small, makes a crucial difference to family well-being. For example, Pahl (1983) found that, as far as household spending was concerned, men contributed more in absolute terms but women contributed more in relative terms: 'Put simply, if a pound entered the household economy through the mother's hands more of it would be spent on food for the family than would be the case if the pound had been brought into the household by the father' (cited in Oppenheim, 1993, p. 109).

As discussed earlier, the presence of children in a family greatly increases the likelihood of poverty. It is a time when expenditure peaks but income falls. The problem of poverty is particularly significant for people bringing up children alone. There are now over 1.4 million lone parents, nine out of ten of whom are women. Over the past decade it has become more difficult for lone mothers to work, either because of the lack of suitable employment or because of a lack of affordable childcare. Consequently while women generally have been participating more in the labour market the participation of lone parents has fallen. As a result, in 1992 nearly 1.1 million lone parents were dependent on income support, 658,000 for over two years.

For the past ten years there has been increasing hostility towards lone parents and the trend in government policies has been away from the problem of poverty among lone parents and their children, to a concentration on the 'problem' of lone parents per se. The government's main initiative with regard to lone parents has been the creation of the Child Support Agency (CSA). While first floated in a green paper entitled Children Come First (1990), meeting the needs of children has in practice come a very poor second to the goal of reducing dependency on welfare benefits. In the early years the CSA was concerned only to collect maintenance from the absent partners (overwhelmingly fathers) of women claiming state benefits. Attempts to secure maintenance from the absent parent were sometimes terminated if the woman came off benefit. The impact of the CSA on absent parents in terms of the additional financial burden placed upon men and sometimes their second families received very considerable publicity which, together with stinging criticism from the House of Commons Social Security Select Committee, led to significant changes both in CSA policy and administration. The fact that the CSA raised only £15 million in new money in its first year, much of which would have been deducted from the benefit of lone parent social security claimants, received little attention. In the 1995 Budget a commitment was made to end preferential help for lone parents. The first move in this direction is the freezing of lone parent benefit and the one-parent premium paid on income support. The poverty which is a common feature of many lone parents' lives can only get worse under such policies.

Although increased longevity among men and women should be a cause for celebration, numerous government reports have instead presented it as a burden: on social security and on the health and social services. The growing affluence of some older people, mainly middle class and newly retired, has informed much political debate and some government policy in recent years. However, alongside the WOOPIES (well-off older people) is a significant number of less well-off, older people, the majority of whom are women living in poverty. In 1994, single pensioners accounted for 11 per cent and pensioner couples 10 per cent of those living on incomes below half average income (Oppenheim and Harker, 1996, p. 35). Poverty among older people like poverty among the general population does not fall evenly: 'Older people comprise a large proportion of those living in poverty in Britain .... But poverty is not evenly distributed among older people, and gender is one of the clearest lines along which the economic and social experience of old age is divided' (Walker, 1993, p. 176).

There are more than twice as many older women as older men living in poverty or on its margins; and among the very old (those over 80) the ratio is five to one (Walker, 1993, p.176). Women's greater longevity over men only partly explains their greater vulnerability to poverty as everyone's income gradually declines when they live longer in retirement. More significant are the social and economic disadvantages they experience during their working lives which are translated into continuing inequality and more financial insecurity in retirement. Fewer working years, more part-time work and lower wages all contribute to poorer pension protection. The Social Security Act 1986 substantially weakened the State Earnings Related Pension Scheme, as a result of which women in particular will lose out. The universal state pension is under threat from both major political parties. The growing emphasis on the need for a second pension, occupational or personal, disadvantages women. They are much less likely than men to have such cover and where they do it is generally less generous (Groves, 1993).

WOMEN AND SOCIAL SECURITY

The British social security system now formally offers women equality of treatment with men. Some of the most unjust inequities, such as the housewives' non-contributory invalidity pension, which required married women to undertake a 'household duties' test in order to qualify and the rule which excluded married women (the largest group of carers) from entitlement to the Invalid Care Allowance have been scrapped following rulings by the European Court.

With the increase in poverty in Britain, the social security system is playing an increasingly important part not only in providing incomes to those outside the labour market but also by subsidising those in low-paid work through family credit and housing benefit. More women than men receive social security benefits. As Table 3.2 shows women predominate amongst claimants of most benefits, with the exception of national insurance benefits paid to people below retirement age. In

Table 3.2 Proportion of recipients of different social security benefits by gender

	Men	Women
Industrial death benefits	Nil	100
Industrial injury disablement benefit	89	11
Invalidity benefit	76	24
Retirement pension	35	65
Widows benefits	N/A	100
Maternity benefit	N/A	100
Sickness benefit	74	26
Unemployment benefit	68	32
Attendance allowance	37	63
Child benefit	2	98*
One-parent benefits	9	91
Invalid care allowance	18	82
Mobility allowance	52	48
Severe disablement allowance	40	60
Family credit	1	99*
Income support	43	57**

* In the case of couples, claims should be made by the woman
** Couples can choose which partner should be the claimant
Source: Lister, 1992, Table 4.1

addition to those who receive benefit in their own right, Esam and Berthoud (1991) have estimated that 3.7 million men receive benefit on behalf of their partner. The contributory benefits system serves women poorly despite the introduction of formal equal treatment rules which give men and women equal rights and entitlement (with the exception of maternity benefits and benefits for widows which are paid only to women and not to men). Entitlement to a contributory benefits system is contingent upon a person's relationship with the labour market. Specifically a worker must earn over the lower earnings limit (at which stage she and her employer begin to pay national insurance contributions) and then must have paid sufficient contributions within the relevant period. It is difficult accurately to assess the numbers affected. However, Hakim (1989) found that the number and proportion of the workforce falling outside the national insurance net gradually increased during the 1980s. She estimated that in 1987–88 about 2.2 million people were excluded, of whom four-fifths were women and over four-fifths were part-time. Overall, women were seven times more likely than men to be outside the national insurance net (Hakim, 1989 quoted in Glendinning and Millar, 1992, p. 136). Changes in entitlement to national insurance benefits, including the extension of the qualifying period for unemployment benefit from one to two years and the abolition of the reduced rate of unemployment and other NI benefits, particularly affected women claimants.

The presence of dependants can lead to women being denied unemployment benefits even where they meet the contributory conditions. As part of the availability for work test, all carers must show that they can make alternative care arrangements within twenty-four hours. Although caring responsibilities can be given as 'good cause' for refusing a job offer, doing so could mean a woman is deemed unavailable for work 'in particular if their availability is being assessed by someone who believes that they should not be in the labour market anyway' (Lister, 1992, p. 33). Under the provisions of the 1989 Social Security Act unemployment benefit is only paid to women if they are prepared to take full-time work (Callender, 1992, p. 136). Thus even those with a full contributions record will be excluded should they wish or be able only to work part-time.

As well as being less likely to meet the conditions laid down in a contributory system, women are less well-protected by the national insurance system because it offers workers protection only against work-related risks, such as unemployment. It does not protect workers against cessation of work as a result of caring responsibilities (Harker, 1996). It has been estimated that only six out of ten women qualify for maternity benefits (McRae and Daniel, 1991). The one benefit paid to carers, the majority of whom are women, is non-contributory but paid only to people looking after someone who is highly dependent and at a lower rate than contributory benefits or income support. The exclusion of many women from the contributory system, together with the inadequate level of national insurance benefits which affects men and women alike, means that an increasing number of women of all ages are dependent on the major means-tested benefits (either as the claimant or the partner of a claimant) such as income support and family credit. However, the nature of means-tested benefits and in particular the aggregation of income of both partners in a couple means that many women excluded from the national insurance system will be ineligible for means-tested benefit because of the resources of their partner.

Nevertheless as Table 3.2 shows, there are more female than male claimants of income support, in addition to which many women receive benefit as the adult dependant of their partner. Since the introduction of income support in 1988, either partner in a couple can be the claimant. However, though there has been a steady increase in the number of women claiming on behalf of the couple, by 1990 only 1 in 20 women had taken the claimant role (Lister, 1992, p. 42). The man is far more likely to be the claimant because he is most likely to have been working full-time and to have had the higher earnings (Millar, 1989; Kempson, Bryson and Rowlingson, 1994). Also in many families there is great resistance to women taking on the claimant role (McKee and Bell, 1985) even though they generally are the domestic managers.

Despite the inadequacy of the contributory national insurance scheme for women in particular, the shift towards means-tested benefits further penalises them. First, as the majority beneficiaries of such benefits, as lone parents, as carers, as older women, or as the domestic manager within a couple, it is they who are confronted with the complexities and indignities of such a system. Payment of income support is only made after the completion of a 24-page application form and any subsequent changes of circumstances must be reported immediately to the Benefits Agency, which administers the scheme.

Secondly, the emphasis within the benefits system on the traditional, though now minority, household structure of a married couple with children, leads to many complexities and inequities. This can have a disincentive effect on women partners going out to work and on lone parents forming new relationships. Both contributory and non-contributory benefits are paid on an individual basis and the earnings of a partner do not affect the claimant's entitlement. However, means-tested benefits aggregate income, so if one partner works over 16 hours per week (the current definition of full-time work in the benefits system) neither partner can claim income support; if either or both partners work part-time or have savings their benefit is reduced. The aggregation of earnings of married and cohabiting couples provides a financial disincentive on top of the considerable social pressures on the partners of unemployed men to stop work. The DHSS Cohort study found that the wives of unemployed men receiving means-tested supplementary benefit were three times more likely to give up work when their partners registered as unemployed than the wives of men claiming contributory unemployment benefit (Moylan, Millar and Davies, 1984, p. 129). This was attributed by the researchers to 'the disincentive effect on the economic activity of wives of unemployed men created by the social security earnings rules' (Moylan, Millar and Davies, 1984, p. 129). A study by the Institute for Fiscal Studies found that receipt of (then) supplementary benefit was 'fundamentally incompatible with women's work' (quoted in Brown, 1989, p. 76).

The social security system has been criticised for encouraging women to leave relationships to bring up their children alone and at the expense of the state (Morgan, 1995). It is true that the benefits system does provide women with an escape route from private dependency on the family and in some cases provides them for the first time with an independent and secure income. Whether they would choose to do this irresponsibly given the low levels of benefits is more questionable. Lone parents are one of the only groups of claimants who report that they are better off when they come onto income support. This is usually because they receive and control the family income directly. Pahl found that between one-fifth and one-third of women in her study were better off once they had left their husbands (Pahl, 1983). The following quote illustrates the change (Walker, 1993): 'I'm better off because I have a set amount of money coming in each week. My husband boozed it.' As Lister points out, the knowledge that benefit will arrive regularly and the security that knowledge provides is as important to lone parents as the actual amount of money received.

> Even dependency on means-tested income support, with all its inadequacies and indignities, can for some women be preferable to dependence on an individual man. It may not be totally reliable; the rights it gives are being eroded; and claiming it can be a humiliating process (especially for black people); but at least the relationship between the State and the woman claimant is a more impersonal one and provides some enforceable rights and some sense of control over the money once received
>
> (Lister, 1989, p. 13).

WOMEN'S EXPERIENCE OF POVERTY

In poor families, especially those on social security (Kempson, Bryson and Rowlingson, 1994) the financial and emotional burden of managing the family budget most commonly falls to women, in contrast to more affluent families where it tends to be a male or shared task.

> ... for women poverty is a different experience, in that it draws on their skills as a manager, a 'good housekeeper', and exacts a price in terms of stress, anxiety and sheer hard work .... 'Managing' also creates an emotional burden — the worry of finding goods, entertaining children with no toys, keeping their minds off their hunger, taking them with you to the shops, and feeling guilty about everything. (Payne, 1991, p. 35)

As the managers in the household, it is the women who not only bear the emotional burden of managing the family poverty but who also make the biggest personal sacrifices: going without or eating cheaper meals (Burghes, 1981; Payne, 1991; Vogler, 1989), withdrawing from social activities and sacrificing personal expenditure in favour of communal consumption (Callender, 1992). Bradshaw and Holmes (1989) found that only 3 per cent of their women respondents went out for a drink occasionally compared to 18 per cent of the men they interviewed: 'By impoverishing themselves women help to prevent or reduce poverty for other members of their family. Thus a woman can be in poverty while other members of her family are not or she may be in deeper poverty than they are' (Oppenheim, 1993, p. 198).

Managing on a low fixed budget takes very great skill and no little sacrifice:

> There is little doubt that the families who were best able to make ends meet were all highly determined managers .... How these determined managers achieve this is by a mix of resourcefulness, ruthlessness and precision. Resourcefulness, because they use their money or time with great expediency; ruthlessness, because they will cut expenditure as savagely as is needed to keep out of debt; and precision because they plan their budgets and household accounts almost to the last penny.
> (Ritchie, 1990, p. 30)

Such discipline and self-determination can be extremely draining for the person with the responsibility for 'managing', as the following quotes from women on benefit illustrate (Walker, 1993, pp. 83 and 81):

> I cope but it's a struggle and my nerves are affected with constantly having to work it all out properly.
> Just really feel alone, isolated, don't know who to turn to for advice ... seem to be hitting my head against a brick wall and getting nowhere. Don't think anyone fully understands unless they've been in the situation themselves. No one seems to care.

CONCLUSION

There have been a number of significant changes in the situation of women in Britain in the period since the Second World War. More have gone out to work and withdrawal from the labour market, which has become contingent on childbirth rather than marriage, has tended to be for shorter periods. Women live longer and they continue to live longer than men. A number of factors have led to an increase in the number of lone parents, including an increasing divorce rate and less stigma and fear of lone parenthood itself. However, these fundamental shifts in the economic, demographic and social position of women have not been reflected adequately either in labour market policy or in the structure and delivery of social security provision. As a result, many women remain marginalised in the workforce (see Chapter 2), concentrated in 'female' jobs, in part-time jobs and in low-paid jobs. The benefits system, while superficially offering women equal treatment with men, is based on the breadwinner/adult dependant model, with the former being male and the latter female. Social attitudes have also been slow to reflect the changing aspirations of many women or the realities of women's lives. For example, many women choose to work as a way of fulfilling their own potential as an individual. But many women have to work out of economic necessity not for 'pin' money and not just for luxuries, but to keep their family out of poverty. The inequality faced by all women relative to men, is exacerbated for some by their social class, age, disability and 'race'. In such circumstances they are further marginalised from the workforce and face even greater economic insecurity. Neither employers nor the state have sufficiently addressed the needs and rights of women either in wages policies, social protection policies or childcare provision.

The evidence on inner-city deprivation is that it is social not spatial: 'we find urban poverty wherever we find the urban poor' (Goodwin, 1995, p. 68). But living in the inner city presents special problems to the poor, and to poor women in particular. The inner city has borne the brunt of major social and economic change: the exodus of manufacturing employment, the decline in the quality and quantity of social housing, cuts in public services and rising crime. Despite the introduction of numerous urban and regional policies to tackle inner-city deprivation and promote regeneration, 'the gap between conditions and opportunities in deprived areas and other places . . . remains as wide as it was a decade ago' (Willmott and Hutchinson, 1992).

The problems for women living in the city remain particularly acute. Their cash impoverishment is exacerbated by service poverty (Alcock, 1993). Cuts in public services, which have disproportionately hit metropolitan areas in the last fifteen years, particularly hurt women. It is they, both as service users and carers, who feel most acutely the lack of safe play space for children, cuts in free or subsidised childcare, increases in the cost of school meals and cuts in social services provision, such as home helps.

Many other factors tackled elsewhere in this volume adversely affect women and serve to increase their poverty, isolation and exclusion: inadequate or expensive transport (see Chapter. 9), higher cost of living as supermarkets move from city centres, fear of crime and for personal safety (see Chapter. 8), and their disadvantaged access to decent housing (see Chapter. 4).

Amid such adversity, there are some encouraging signs. Philo (1995), for example, points to numerous community initiatives aimed at tackling some of these problems and the prominent role played by women in these initiatives: the establishment of credit unions, food co-operatives, childcare and after school initiatives. Morrell reports on one such project in Chapter 20.

In the longer term, however, the solutions to women's poverty discussed in this chapter can only be overcome by structural change, which in itself may need a shift in attitude by both men and women. Genuine equal opportunities for women in work, supported by adequate childcare provision and social security policies which both acknowledge women's dual role as workers and carers and which recognise their rights as individuals not only as the dependant of a man. This in turn requires governments to turn their attention to recognising the problem of poverty — for all — instead, as in the recent past, of trying only to blame and marginalise the victims of poverty, the poor.

SECTION 2
Women and Home

4.

HOUSEHUNTING

Jane Darke

INTRODUCTION

This chapter looks at the ways in which women may obtain, or fail to obtain, somewhere to live. It analyses the problems and barriers faced in gaining access to a satisfactory home and describes how women's actions have improved the livability of some areas. It will examine how and why some women face extra difficulties in obtaining a home. The chapter looks at difficulties arising from women's economic situation and their choices about lifestyles. The next chapter will consider another aspect of women's housing situation: gender differences in the relationship to home and the meanings assigned to home, and how women's experience of home is affected by domestic tasks, which are still seen as women's responsibility rather than men's. The search for a home is presented as a househunt considering each tenure in turn, supported by some particular examples: the losers in the housing lottery, and the problem of relationship breakdown. Finally, the chapter looks at some positive examples and possible changes to the system.

Women's housing needs are diverse. Living with another adult may or may not alter a woman's housing preferences, but if both adults are earning, it will make those preferences more attainable. Having children radically alters a woman's housing priorities and continues to do so as the children's mobility and spatial range increases and their interests change. Different ethnic groups will vary in the attributes of a home that they value, and there are locational choices which may involve proximity to (or avoidance of) those of similar backgrounds and, for some, the need for collective resistance to racism. For those who are disabled, the detailed design of their housing and its location may be critical to their quality of life. Women's housing opportunities follow from their choices of who they live with, while their life choices such as careers, marriage, cohabitation and motherhood, may be influenced by the consequences of those choices for their housing opportunities.

There are two types of reason for women's disadvantaged access to housing: economic and ideological. Women earn less than men and hold less wealth (see Chapters 2 and 3) but they are also affected by norms

48

about 'proper' living arrangements. The options on how and with whom to live are constrained by the social disapproval that may follow some choices. Over half of all households currently consist of couples with or without children. This is due to fall to less than 50 per cent early in the next century (DoE H&CS) but the housing on offer still reflects the view that this is the only legitimate household form. Over the last few decades, both the housing situation and the nature of the household have undergone dramatic changes. Table 4.1 summarises some of them.

The social changes can be summarised as a move away from the stereotyped nuclear family towards a greater diversity in household forms. The increases in single-person and single-parent households are documented in official statistics, and are forecast to continue at least until 2011 (DoE H&CS 1984–94, Table 9.10). The majority of both these types of household are headed by women. Other changes are not so clear: for example the census coding system does not accept that same-sex sharers can be 'living together as a couple' (Heath and Dale, 1994, p. 7): lesbian and gay couples are literally invisible in official statistics. The statistics also fail to show couples who live apart some of the time, perhaps for job-related reasons, children who spend part of their time with each parent in different households, households that include step-relationships, communes, travellers or squats, all of which are part of a new diversity in household forms. Meanwhile, the government's statement of its future directions in housing policy speaks of 'the need to support married couples who take a responsible approach to family life' (Cm 2901, 1995, p. 36). Where have the policy-makers been?

Not only the government but also most housing providers still think in terms of family households, and most social landlords give priority to those with children. It is as if Britain was still in the 1950s, when the post-war housing crisis meant that single women had to live in somebody else's home: preferably their parents' until the almost obligatory marriage and possibly after; with grown-up offspring or other relatives for widows, as a lodger with a family for women working away from the parental home. The high proportion of private tenants

Table 4.1: Change in housing situation and lifestyles, 1951–3 to 1991–4

	1951–3		1991–4	
Owner-occupier households	32%	(1953)	68%	(1994)
Renting households (all sectors)	68%	(1953)	32%	(1994)
Households lacking or sharing bath	45%	(1951)	1%	(1991)
Households sharing with others	29%	(1951)	1%	(1991)
Households of one person	1.4m	(1951)	5.9m	(1991)
Births outside marriage	<5%	(1951–3)	32%	(1993)
Divorces (thousands)	30	(1951)	175	(1992)

Sources: Holmans, 1987, Tables V.1, IV.15 and IV.12; OPCS Censuses of 1951 and 1991, various tables; OPCS Birth Statistics, various dates; CSO 1995a, Tables 2.18, 2.22 and p. 42.

did not mean that mobility was easy or that single people had opportunities for living independently: housing was often allocated by being 'spoken for' by an existing tenant with a family member needing a home (Young and Willmott, 1957, Chapter 2). The council houses being built in large numbers (a record 230,000 in 1953: see Merrett, 1979, appendix 2) were almost all for families, selected for their respectability, stability and their ability to pay their rent regularly. It was common practice for housing managers to check an applicant's marriage certificate before offering a tenancy.

Our househunt will start by examining owner occupation, move on to look at social housing (renting from councils or housing associations), then look at other housing situations. In Britain, the advantages, reputations and problems of different dwelling places are strongly linked to tenure, although in most of Europe housing tenure is not seen as a component of status. Home ownership in Britain is seen as having higher status than renting but the possibility of attaining this tenure is related to economic position. Over the past few years, the development of a hierarchy of desirability of tenures has intensified. There have always been those who looked down on council tenants and treated owner occupation as a badge of status, but the significance of different tenures has changed. In the 1950s and early 1960s there were more households in private than council renting, and very little high-rise housing had been built. Many women in housing rented from private landlords were impatient to get out of very poor quality accommodation, lacking baths and inside toilets, let alone hot water and central heating systems. Council housing was seen as an enormous improvement on what was on offer from private landlords, and council tenants were similar in income and social class profile to the national average (Willmott and Murie, 1988). The high-rise building of the 1960s, although never more than a small proportion of output, changed perceptions of council housing and stamped the sector with a disastrous image which has enabled criticism to become a self-fulfilling prophesy.

A HOME OF YOUR OWN?

Current housing policy assumes that owner occupation is everyone's preferred tenure and that if you can afford it, it will be your choice. A preference for renting is seen as eccentric, a choice that has to be accounted for, in contrast to the 'naturalness' of home ownership (Gurney, 1995b). Certainly, ownership gives greater choice for those who can afford it, even marginal buyers.

At the time of writing, in late 1995, first-time buyers in southern England would have to spend between £40,000 and £50,000 (unless they settled for a 'park home', a euphemism for a mobile home on a permanent site) and could buy a house for a few thousand less in cheaper parts of Britain. It is often said that there is little choice for first-time buyers, but within this price range they can choose between a three-bedroom, ex-council house (on a 'less desirable' estate), a studio flat in a sought-after location, a new two-bed starter home in a suburb or small town, an older terraced house needing some renovation, and several one- or two-bedroomed flats. Compare that with the procedure in social renting, assuming that a household qualified for this: view one property at a time on a 'take-it-or-leave-it' basis, with an uncertain wait for the next offer and a possible loss of priority in the event of a refusal. The advantages of ownership are real and it is hard to see how procedures in social housing management could alter to offer a similar degree of choice to that of the private sector: choice is something markets do well.

But who can afford to buy? Most households with two or more earners can, but very few can with one earner or none. The average price paid by first-time buyers in early 1995 was £46,000, with an average income for borrowers of just over £18,000 (DoE H&CS, March 1995). Only 22 per cent of women in full-time work earn this much compared with 43 per cent of men (CSO, 1995b). Furthermore, only just over a third of women work full-time compared to three-quarters of men (excluding those over retirement age). Clearly, home ownership is most affordable for couples or sharers, then single men, and finally single women, with perhaps one in thirteen of them being able to afford to buy in their own right.

So a woman's best route into home ownership may be through a relationship with a man in secure and/or well-paid employment (Gilroy, 1994). Even then she may not live happily ever after: about three million households are, at the time of writing in late 1995, either in negative equity or have insufficient equity to move (ROOF Briefing, 1995). This is almost one in three of those with mortgages. Worst affected are: first-time buyers, those who bought in 1988 or later, buyers in their twenties and buyers in southern England (Dorling and Cornford, 1994). About a thousand households a week are losing their homes through repossession. There are also the problems of housing following relationship breakdown, discussed below.

It has taken time to unlearn the lessons of the mid-1980s when it was possible to make more money from house price inflation than from a job. The changes in the housing market particularly affect women. It has become almost impossible for recent buyers to move on, and difficult to step off the treadmill for long enough to have a baby. It may be just feasible if you can guarantee a problem-free pregnancy, full-term delivery of a single healthy baby, cheap childcare from family or friends, and a sympathetic employer who recognises that parenthood involves time commitments. The absence of any one of these can spell disaster for the home owner. There is evidence that women are waiting longer to start a family and that families are getting smaller, while some women who have waited may discover that they have missed their opportunity. The market that offers a wide choice of property has its downside: a high risk of involuntary childlessness, repossession or being stuck in a tiny starter home with a growing family.

SOCIAL HOUSING

If it is difficult for women to attain the prize of home ownership, do they then have priority in access to social housing, as is widely believed to be the case, especially when it comes to single mothers? The proportions in social housing are: 17 per cent for married couple households, 30 per cent for single men and 40 per cent for households headed by women (OPCS, 1995, Table 3.10). Comparing one- and two-parent households, 57 per cent of the former are in social housing compared to 17 per cent of two-parent families (ibid, Table 2.26). In fact all households must be assessed on the same basis under the equal opportunities legislation, so that if women are over-represented in social housing it is because of their greater need. Need is the currency of distribution of social housing just as cash and credit-worthiness is for owner occupation.

The term 'social housing' includes council housing and housing associations. The latter are non-profit organisations that build for rent (or for shared ownership: part rent, part buy). About half of the capital funding for their housebuilding comes from government via the Housing Corporation, with the rest from private borrowing. Rents tend to be slightly higher than those of local authorities but like local authorities they allocate their homes on the basis of social need.

Local housing authorities and housing associations may devise their own allocation schemes within guidelines set by the government, which include priority for those in poor conditions or who are overcrowded. There are two main routes into social housing: via the waiting list or through homelessness. Chapter 6 explains the various tests which a household must pass before they can be rehoused as homeless. In theory it is easy to join a waiting list, and it has long been regarded as best practice to allow anyone to register (CHAC, 1969) although many local councils set a minimum age (for example, 55) below which single people will not be considered unless they are vulnerable (Venn, 1985). Once on the list, some cases must receive higher priority than others.

Issues about how social housing is prioritised have become much more salient because of the overall reduction in the quantity of social housing available to be allocated. Ever since the late 1970s councils have had their building programmes cut and are now virtually debarred from building new homes. Their existing stock has been reduced through disposals under the right to buy scheme; although this has now slowed to a trickle, it is the best quality stock that has been lost. At the end of 1979 there were 5.2 million council homes in England; at the end of 1994 there were 3.7 million. This loss of 1.5 million has been only partly made good by an increase in housing association homes of half-a-million (DoE H&CS, various dates; figures are given for England only because tenure breakdown for other parts of Britain is less detailed). Housing associations have become the main providers of new social housing but an increasing proportion of their vacancies are allocated through nominations from local authorities.

Most councils set priorities through a system of allocating points for different needs but this cannot be considered a scientific measure of need. How would you set priorities as between a young mother who is very keen to move out of her parents' home, a lesbian couple who are being harassed, two friends with learning difficulties who want to try living in a home of their own, and two elderly sisters who cannot afford to repair the damp house they have inherited? To which would you offer the only two-bed house that has become available this week? An allocation system is best understood as a set of decision-making rules for housing workers to protect them from constant canvassing or criticism, rather than as an exact science. The actual likelihood of rehousing depends on the area: in some London boroughs it is negligible however high you are on the priority list, because almost all vacancies go to homeless households. Only in a few areas which are losing population and jobs can you expect quick rehousing. The problem is the shortage, not any particular system of allocation.

It would be illegal as well as unjustifiable to take the applicant's gender into account in setting priorities. Women have priority only as carers of dependants. Women living alone have poor prospects of obtaining council housing unless they are elderly. Is there any evidence of discrimination *against* women in social housing? Why are so many applicants allocated housing that is not what they would choose: a flat rather than a house, an estate where people fear crime, an area of racial tension? Why do many existing tenants believe that their landlord has given up on their estate and is only rehousing people with problems?

The fact is that social indicators for the entire social housing sector have been worsening as one facet of the increasing social polarisation of the 1980s and 1990s. The gap between rich and poor has widened and the poorest groups have become more concentrated in social housing (Willmott and Murie, 1988). The average income of tenants of council housing is only 48 per cent of the average across all tenures (OPCS, 1995b, Table 3.15). Among new tenants of housing associations, unemployment is over 50 per cent, and over half the families with children moving into council housing are headed by a lone parent (Page, 1994). Many will have suffered the trauma and disruption of homelessness. Others must be rehoused because mental illness, or a history of crime or abuse makes them vulnerable. In conditions of extreme shortage, it is inevitable that such households will have priority over those with fewer problems. However, the consequence is that some housing areas have such a concentration of people trying to manage difficult lives that there are not enough people with the energy and capacity to help the community to function as it should.

It is vital that the capacity of a community to sustain social life is considered both in national housing policy and in local decisions on how particular estates are allocated. For example, in Sheffield, residents on a large block of council flats noticed that many of their new neighbours had moved from a large mental hospital which was due for closure. They got organised: not to protest against these neighbours but to campaign to have support facilities provided for those who needed them. Too large a concentration of people with problems would have meant not only a loss of goodwill but a swamping of the capacity of estate residents to work together for the benefit of the whole community.

In the areas where turnover is greatest, where people go to great lengths to move out, the concentration of people with difficulties in their lives will increase more quickly. It will be even more acute on a newly built estate let for the first time. At the same time, the lives of the poorest 20 per cent have become more difficult to manage with higher unemployment, lower benefits (especially for young people), the ending of single payments for the cost of setting up house, and cuts in council support in areas such as nurseries, cheap fares, home helps, youth work. In this way the poor reputation of social housing has become intensified.

The reaction of some residents, especially young men, has served to further worsen the social environment. Joyriding, breaking and entering, racially motivated violence, drug dealing or merely absenteeism from child-rearing contribute to the breakdown of community life that Bea Campbell (1993) has analysed. Where residents are organising to improve their areas, women are invariably in significant roles. The hope for run-down areas lies in women's co-operation, but this cannot succeed in the long term without changes to the housing system as a whole designed to create better balance. A community needs skills in which women excel: negotiating, conciliation, persistence, neighbourliness, explaining a problem in writing, understanding the benefits system.

WHO LOSES OUT?

Some groups face extra problems in gaining satisfactory housing. The housing system is geared to the expectations of the predominant group: able-bodied white British. The norms of what constitutes a 'proper' home, sometimes including its tenure, are culture-specific. Those who migrate to another setting have little choice but to adjust to what is available in the new country. Even their children and grandchildren cannot be assumed to have converged to the dominant cultural norm, which itself changes over time.

Black households buying a house may be given different information or 'steered' to particular areas by estate agents and mortgage lenders (CRE, 1985, 1988) and are always conscious of the risk of racist reactions from neighbours. In social housing as well as for estate agents and lenders, it is illegal to discriminate on grounds of ethnic origin. Most social landlords have formal equal opportunities policies yet every investigation has shown unequal treatment, with black households getting poorer quality housing (see for example CRE, 1984, 1989; Henderson and Karn, 1987; Jeffers and Hoggett, 1995). Some of this is due to prejudices held by housing officers but there are other factors at work that should be researched: procedures, the mix of properties in the stock, stereotyping, applicant strategies, racism on some estates, differential urgency of housing need, and so on (see Jeffers and Hoggett, 1995). Racial discrimination is also widespread in private renting, to be discussed below (see CRE, 1990).

There are specific issues for those who are disabled. Firstly, most houses are not designed to meet their needs. Secondly, disabled people often have reduced earning capacity and face job discrimination, so are less able to meet their needs through the market. Their requirements may be more precise: accessibility of facilities, proximity to those who provide support, for example. Although the market fails to deliver, social housing too marginalises disabled people by applying the label 'special needs'. Many disabled people have to fight against the insensitive but well-intentioned stereotyping that denies them the right to independent living, assuming they should be 'cared for' by others. The importance of the home for personal autonomy is further discussed in Chapter 5.

There are further factors which give rise to housing disadvantage (for example age, at both ends of the housing career) though it is not necessary to catalogue them. All these groups would benefit from a change in the present system which treats one tenure as 'normal' and 'natural' and all others as catering for the failures of a market system.

PRIVATE RENTING, HOSTELS AND FOYERS

What are the alternatives for those, including many women living alone who have neither the market position to attain home ownership, nor the currency of need for access to social housing? In most parts of the country the remaining option is to rent privately. This is a poor solution in many ways: privately rented properties are far more likely to be in poor condition than other homes (DoE, 1993) while new lettings are offered at market rents but are only eligible for housing benefit up to the local average rent. The high rents in this sector cause serious difficulties not only for tenants, especially those who have to contribute towards the rent from their benefits, but also for the housing benefit budget.

High rents do not necessarily buy quality: many tenants have to share facilities such as bathrooms and cooking facilities. Shared accommodation may be particularly difficult for women to tolerate. In addition to inconveniences such as having to clean the bath before use, women are at risk of sexual harassment or assault from fellow tenants, intruders or the landlord. Behind the front door to the house, rooms are more vulnerable to break-ins. As in other tenures there is little choice of neighbours but in private renting the dividing walls are usually thinner. As most accommodation is let furnished, a tenant must live with someone else's choice of furniture, sometimes of doubtful cleanliness and failing to meet safety standards. There is a shockingly high rate of deaths from faulty gas appliances or fires in houses where the landlord has disregarded safety regulations. This is a tenure of transience; people moving jobs, moving houses, between relationships, on short courses or placements, or trying to leave sleeping rough. Most furnished tenants stay less than a year (OPCS, 1995b, Table 3.12).

Single women in the past had other options: for example many large employers used to provide staff hostels (Dodwell, 1995). Some rooms were shared, men were not admitted except in defined parts of the building, and there were rules about how late residents could stay out. These conditions seem to have been accepted as being in the employer's reasonable interests although they would not be tolerated today. This option no longer exists but there are some hostels for single people (far fewer for women than men: see Austerberry and Watson, 1983) and a new form of provision in some areas: foyers. Foyers are aimed at young people and combine hostel accommodation with preparation for employment. They have been an established part of housing provision in France for most of the post-war era, and often include café and drop-in facilities. The idea was adopted in Britain in the early 1990s. Supported by an unlikely alliance of the housing minister with the pressure group Shelter and some business interests, a handful of foyers had opened by the end of 1995, some purpose-built and others adapted from existing hostels.

Although some foyers seem to work well there have been a number of problems in achieving a balance between the housing and training objectives. Should they be used for housing young people with good prospects of moving off the unemployment register or those in greatest housing need, who may be coping with personal problems that make it difficult for them to get and keep a job? A resident who is not thought to be actively seeking work or training could be evicted, as could someone on maternity leave. Training in how to manage a home or a tenancy for residents who may be living independently for the first time is not necessarily provided, while staffing costs and hence rents may be very high, placing residents in a poverty trap and a burden on public expenditure (Anderson and Quilgars, 1995). As a way of addressing the housing problems of young people they are too expensive a solution to be provided in large numbers. Young people are particularly disadvantaged in a housing system that relies mainly on the market (Darke *et al.*, 1992).

RELATIONSHIP BREAKDOWN

Our housing needs change over time. Most of us start our housing careers in poor quality accommodation but are able to improve our situation over time, at least until old age brings further changes in our needs whilst reducing our capacity to provide for them. A significant reverse in the upward trend is caused by relationship breakdown. The housing system is not geared to family break up and may contribute to worsening the relationship between couples who split up and between children and the non-custodial parent.

The situation of couples who split up differs according to their tenure, the circumstances of their break up and whether or not there has been violence. For owner-occupiers, even if both parties are earning this may not be enough to enable them to buy separately, for the reasons outlined earlier. It is a myth that women get to keep the house. The proportion of owner-occupiers among divorced and separated household heads is the same for men and women: 49 per cent in each case, compared with 78 per cent for couples (OPCS, 1995, Table 3.10). The proportion of women who keep the family home may decline in future. In the past the transfer of the home may have released the ex-husband from a possible court order for maintenance, but under the Child Support Act 1991, child maintenance must be paid in any case. Often an ex-wife has the right to remain in the matrimonial home for only as long as there are dependent children. As soon as her children grow up and leave she may become homeless, with no priority for local authority rehousing. Furthermore, she may be unable to move house even before this, as the

cost of legal aid may be the first charge on the proceeds of sale, making
the move unaffordable.

In housing terms it may be advantageous for the man to walk out
rather than for her to leave: at least she is left in the house, albeit in
emotional and financial turmoil. Where a woman leaves this may or
may not be due to violence from the husband. In theory, domestic
violence is reason for being given priority as a homeless household but
there are barriers to be overcome first. Violence is disempowering; it can
be very difficult for a battered wife to take control of her situation by
walking out (Dobash and Dobash, 1980; Binney et al., 1981). There is a
real risk that the very attempt to leave may provoke further possibly
fatal violence. If a woman does move she may find it difficult,
distressing or humiliating to explain her situation to a housing official
who sees his or her role as to exclude fraudulent applicants. The
situation of homeless women is described further in Chapter 6.

The Women's Aid movement started in the 1970s as a way of helping
women to escape from domestic violence by providing refuges.
Although there have been differences in management styles (see Dobash
and Dobash, 1992, Chapter 2) refuges typically offer an open door to
any abused woman and are democratically and collectively run in a style
aimed at supporting and empowering residents. Their locations are
confidential, with good reason: murders by outraged partners have
occurred where addresses became known. Refuge workers will help the
woman resolve her housing problems by means of an injunction against
the violent partner, or rehousing — which may be in another district if
requested, as the provisions in the homelessness legislation requiring a
'local connection' do not apply where the applicant is at risk of violence.

Women's Aid groups in most cities have provided a valuable service,
but suffer from insufficient accommodation, underfunding and sometimes a lack of co-operation from housing departments (Binney et al.,
1981; Welsh Women's Aid, 1986). Male policy-makers and liberal
feminists have found it hard to accept the extent of male violence and
some housing officers still fail to recognise the need for rehousing.
Owner-occupiers may be told that they have other solutions available
than taking up social housing. Even if a woman is accepted for
rehousing she may face a long wait and/or rehousing in a type of
accommodation or area she would not have chosen.

For a woman who chooses to leave a relationship without having been
subjected to violence, the situation is difficult in that her need for
rehousing may not be accepted by her social landlord. In theory she can
ask the courts to weigh her claim to the rented matrimonial home as
against that of her husband, but she may be unwilling to compound his
problems by making him homeless (Brailey, 1986). Whichever party
becomes homeless, their prospects are bleak if they do not have care of

dependent children. The lack of suitable accommodation may make it difficult to sustain a relationship with children, if they cannot visit the non-custodial parent's home or if this is an environment which makes visits uncomfortable occasions. Conflicts over the home as well as over children may mean that the non-custodial parent will not make the effort to maintain contact with children.

ALTERNATIVES

The choices for women are difficult. Each form of provision has a problem: high cost, complex entry rules, unattractive housing, poor standards, a difficult social environment, insecurity and intrusive rules. Are there other routes into housing? What might we want to change about existing provision? One possible solution would be to deconstruct the conventional household consisting of an individual or family surrounded by a party wall, to allow more fluidity of grouping. Arrangements other than living alone or with partner and children are currently seen as transitional or second-best, but could have positive attractions. New forms of household, separate some of the time but combined for some purposes, could be created. If, as we are often told, two parents are better than one, might not three or eight be even better? Many individual households cannot afford a large space for a party or gym, or equipment and gadgets for occasional use. Some form of collective or co-operative housing could resolve these dilemmas. Such provision exists in some countries, especially Denmark, Sweden and the USA: see McCamant and Durrett, 1991; Woodward, 1991; Franck, 1994.

Potentially women could work together for their own solutions. In practice there are few instances in Britain of alternative tenures or living arrangements. Options are being reduced rather than expanded: the Criminal Justice Act 1994 further criminalised squatters and those who have chosen to live as travellers. More co-operatives could be set up, but they face a great deal of bureaucracy to obtain funding, and most women do not want to spend a large part of their lives organising to procure a home. Housing associations in Britain will not usually help a group seeking a collective solution unless its members belong to categories who would be on their priority list in any case. New arrangements are almost invariably more difficult to establish than well-tried models. Building societies have belatedly learned to accept cohabitees and sharers but would require a lot of convincing before funding a collective household. Sadly, such households seem to have an even higher propensity to break down than couple households but when they work, can offer real benefits to participants. It would be useful to have further research on collective solutions. Existing studies, now rather old (Eno and Treanor, 1982; Abrams and McCulloch, 1976) have *not* shown that gender roles are changed in these settings, but all-women or women-led groups were not looked at.

Self-build groups are a particular form of co-operative and again it is common for a group to spend years assembling a viable financial package. Many groups fall by the wayside in the process. Self-build is more common in many European countries (see Broome and Richardson, 1995, pp. 113–18, 240–2), some of which have a housing system where land is cheaper and development finance easier to obtain. Although self-build makes a significant overall contribution to housing production in Britain, it could make a far bigger contribution if the housing finance system could come up with a standard package rather than requiring each new group to reinvent the wheel, and if the delivery of training and technical aid could be improved. The author has heard rumours that women's self-build groups exist but there have been virtually no reports on them in mainstream housing or architectural periodicals. Self-build could help many more people to solve their own housing problems, but could never meet the housing aspirations of the majority of households.

CONCLUSIONS

The housing system is geared around a mythical world where everyone has a job sufficiently well paid to give access to home ownership, everyone lives in happy families, and marriages only end through death. It does not cater for the real-world choices that women make, or are forced on them. Not only does the market fail to cater for many situations: the fact that market solutions are seen as 'normal' and all other forms as exceptional, marginalises the growing number who cannot meet their needs through the market. However, social housing as presently constituted is equally incapable of meeting those needs. This failing cannot be corrected merely by increasing resources and the resources in any case are limited. Completely new thinking is needed on how to accommodate the profound changes in lifestyle that have occurred and it must include changes within each tenure, new tenures, new relationships between tenures and more fluid and variable boundaries around the household.

5.

THE ENGLISHWOMAN'S CASTLE, OR, DON'T YOU JUST LOVE BEING IN CONTROL?

Jane Darke

INTRODUCTION

The previous chapter discussed the difficulties women may face in obtaining a home. This chapter examines the significance of home and attempts to explain why the home is so important for women — and to understand the way our feelings about the home are tempered by the work of running a home and of impression management in relation to the home. Women do not share public space equally with men. Chapter 7 contends that the public realm is seen as 'belonging' to men. Women may be tolerated or politely welcomed as guests (or attacked as intruders) but do not control the space. Does this mean that the home is a zone where we *do* belong and are in control?

Women are in a particularly contradictory situation here. Firstly, we need a space where we can 'be ourselves' precisely *because* our behaviour in the public realm is subject to control by others. It has been argued (Saunders, 1990) that there is no difference in the meaning of the home for men and women in that both see it as a haven. Other authors have argued that there are differences; their views will be considered below. Even if the home is seen as a haven for both men and women, what it is a haven *from* may be different: typically for men from the routines of work and for women from patriarchal control in the public realm.

Furthermore, the home is a place of work: it and its occupants have to be maintained to a socially acceptable standard of cleanliness and tidiness. Even if this task is shared with other household members, social expectations will hold the woman responsible for any lapse from the desirable norm, but will also give her the credit for competent performance of this role. The performance may go beyond the merely competent: the home can be a site of creativity, pleasure and self-expression. This is an aspect of the home-as-haven, but it also involves work and public judgments on the homemaker's personality as expressed in the home. Because we are social beings we need to feel valued as members of a society. Although the home may be a haven we do not want total privacy in every part of the home; there must be some connection to the outside world. This may be more important for those *not* in paid work.

There are three components, then, to a woman's relationship with the home. It is a worksite, it is a source of judgments by others about the presentation of the home and its occupants, and it is a haven from an alienating world. Different domestic arrangements and forms of living will affect the balance between these. The presence of children creates work in keeping them and their clothes clean and in teaching them about impression management: putting toys away, not leaving dirty mugs or (later) beer cans around. Some people are reclusive, others keep virtual open house, and in both these extreme cases may or may not be 'houseproud'. The amount of work required in the home increases too with the presence of an adult who requires care. It makes a difference whether the homemaker and her co-residents are at home during the day. An inherently problematic situation is where a partner or adult children are unemployed or homeworkers and the homemaker goes out to work: she is unable to control the presentation of the home during this time. Many women find it hard to adjust to a partner's retirement: this removes valued autonomy and makes it more difficult to maintain relationships with friends (Mason, 1988, 1989). The single householder may have greater freedom to vary the amount and timing of housework. Living alone may mean that more importance is placed on the home-as-haven but both physical work and impression management are almost always necessary.

In all three of these areas the woman householder's control, while not total, is greater than she can exercise in most other spheres of life. We consider first the way the design of the home makes it more or less private, then reflect on the home as worksite and the home as haven. We consider how a woman's control may be compromised in practice: by social expectations, by the design of the home and by the behaviour of other residents and by those who may visit.

IMPRESSION MANAGEMENT AND HOUSING FORM

It is easy to forget how pervasive the social imperative for impression management still is. There are various signs or symbols which may signify that the household is bohemian or belongs to a category seen as less competent in household management. What would the average person in the street assume from the following:

- a front garden that contains litter such as crisp packets and broken glass;
- front windows that have clearly not been cleaned for several months;

- curtains visibly dirty, torn or hanging off their runners; and rooms visible from the street being very untidy?

The list could continue. If we enter such a house, there may be other signifiers of taboos broken: crockery lying around, dirty bedding, an unflushed toilet. Anyone who visits women in their homes without a prior appointment, as a friend, a neighbour or an interviewer, will have noticed how most homes are routinely immaculate, and will also be familiar with the ritual apology for the state of the house: a claim that the speaker aspires to higher standards than those presently evident. It is rarer to hear such an apology from a man. Most women who have ever run a home can recall some embarrassing incident where someone 'found out' that our standards fell short of a norm. Mason (1989, p. 120) refers to her respondents' horror stories about people (usually older men) who allow their standards to fall. The television series *The Young Ones* derived much of its humour from the spectacular breach of the taboo on dirt or disorder in the home.

The design of the home may assist in impression management by the differentiation between front and back. The front-back distinction is found in most cultures but its expression in housing form will vary. Most societies distinguish a public and private realm, usually correlating with male and female spheres and with some degree of male control over female behaviour. Typically, areas used by women, including kitchens, are placed in the back region of a house (Matrix, 1984; Hillier and Hanson, 1984; Chua, 1991). There may be a completely separate circulation network based on back courts and alleys where women meet and exchange news or goods with other women.

Looking at the history of housing for working people in Britain since the nineteenth century, we can see the transition from the crowded, insanitary court to the byelaw street and then to the suburban house. The front-back relationship is different in each case. In the early phases of urbanisation there were few constraints from byelaws or public health regulations, so the form of development was dictated by patterns of finance and landholding and the imperative to maximise the number of lettable units on a site. Living patterns could be chaotic: a jumble of back-to-back cottages around a court that also served as an open sewer, earth-floored, single-room cellars or common lodging houses where families were indiscriminately mixed in with single people of both sexes (Burnett, 1986).

In a court, each dwelling occupied one room in plan, to a height of two or three stories, its windows a few feet from other windows across the access way. A dwelling may have housed lodgers and extended family members in addition to the nuclear family. The boundaries around the single-family household were non-existent at times but a collection of people could share a court almost totally secluded from the gaze of the censorious public. There was privacy for the court but little for the individual household: in a back-to-back there *is* no back region.

The middle-class concerns about the housing conditions of the working class can be seen as an attempt to persuade the working class to adopt the privatised, domesticated lifestyle with clearly articulated gender roles that the middle class had so recently defined for themselves (see Davidoff and Hall, 1987), and to provide the type of dwelling that would encourage this form of behaviour. It remains a great unanswered question how far this change was imposed on the working class and how far it was a way of living to which they actively aspired. Certainly the latter was the case by the time of the First World War (Swenarton, 1981; Ravetz, 1989).

The change in the form of working-class housing from court to street can be traced to the influence of the Public Health Acts of 1848 and 1875, which specified standards of spacing, layout, ventilation and sanitation. The 'byelaw street' was found in most industrial towns although variations in local byelaws meant that each area evolved its own variant (see Muthesius, 1982, Chapter 12). The predominant form was the 'two up, two down' terraced house, each room around 3.2 to 4.2 metres square. Daunton (1983) made the link between house form and family values, contrasting the private house on a street with the 'promiscuous' sharing of semi-private cells of dense development in courts: '[t]he emergence of the private, encapsulated dwelling was a physical demonstration of the social value attached to the conjugal family and domestic life' (p. 37).

How was the small four-room house occupied? The downstairs front room, facing directly on to the street or with only a tiny strip of garden, was regarded as the 'best' room, and was the only one visible to outsiders. Here the highest quality furniture that could be afforded was carefully arranged and kept immaculate. This was the 'front region' by which the family's social standing would be judged. The back room would be the setting for most family activities: food preparation and consumption, childcare, laundry and bathing as well as sitting and relaxing. Using the front room for any of these activities risked undermining its elaborate presentation. It was precisely this lack of use that enabled the parlour to perform its impression management function. The immaculate front room not only asserted to passers-by the skills of the housewife within; it could also be used as a reception room for callers such as the vicar, officials or a potential son-in-law. This house type was extremely well adapted to permit appearances to be maintained with the minimum interference to the normal untidy transactions of family life.

The form of housing changed again with the increased involvement of the state from around the turn of the century, and this was the subject of conflicting opinions. The evangelistic Garden City movement held that houses should be built at lower densities with larger gardens, oriented to receive sunlight even if this failed to respect the norms of front and back. One of the movement's most influential propagandists, Raymond Unwin, advocated a single, large through living-room rather than a multiplicity of small spaces. This would clearly have made the task of impression management more difficult. When a Women's Housing Subcommittee at the time of the First World War systematically canvassed the opinions of working-class women, it was clear that a parlour was wanted (McFarlane, 1984; Ravetz, 1989). Respectable working-class women wanted visibility and recognition for the best examples of their handiwork, together with privacy for 'work in progress', including the ongoing task of adequately socialising the next generation. There is variation in who is permitted to view the 'work in progress': close kin will usually be given more freedom to enter the 'back region' than more distant kin, friends more than neighbours. The front-back distinction is a continuum rather than a dichotomy, with the whole house as a back region compared to the street, but with some parts of the house (such as the bedrooms) further 'back' than others.

The double-fronted house favoured by the garden city movement (Swenarton, 1982, pp. 22–3) exposes more of the life of the household to the public gaze; impression management is harder and high standards must prevail at least throughout the ground floor. The 'through living-room' house continued to be endorsed in government guidance to designers of housing throughout the peak years for council house building (Ministry of Health, 1949; Ministry of Housing and Local Government, 1952, 1968).

Individual households differ in how they set the boundaries between front and back. In a study of households in France, Bernard (1991) distinguished between those which expected visitors to stay in the same room, to use several rooms but not the whole house, or to use the whole dwelling. This varied according to age, social class and occupational category, but there were individual differences within any one category. Thus the boundaries to the back region can vary but it is important to the occupier that she controls these boundaries. Various devices for doing so may be employed: curtaining, leaving doors open or closed, explicit rules — but there is always the possibility that her power to control the situation may be negated.

Housing design today, especially the type of new homes available to the marginal purchaser, continues to favour open planning and the through living-room. This makes the task of impression management even more problematic. Madigan and Munro (1991) show that houses with an open plan on the ground floor, especially those with the single living-space visible from the front door, appear to cater for the household of democratic, egalitarian relationships but in fact require constant work to maintain acceptable levels of cleanliness and tidiness.

On the other hand, the flat is a problematic housing form in that it tends to offer either too little or too much privacy (or even both at once). In a flat there is a lack of aural privacy; quarrels, music and sexual antics are heard without knowing whence they come, creating fears that one's own domestic activities can likewise be heard by others. Some types of flat, such as balcony access, have insufficient visual privacy: neighbours passing a few inches from the window can scarcely fail to notice whether or not the interior is clean and tidy. Other flat types offer no opportunities at all for public display: all the skills and dedication of the person caring for the home go unperceived and unrecognised. Being able to watch the world from above does not necessarily confer a social persona. In neither case is it easy to control the presentation of the home.

In the legendary 'bad old days' in housing management, an inspector from the housing department might call unannounced at any time and would note the cleanliness of bedding and whether there was fluff under the bed. Other figures of authority may invade: families, actually women, are policed (Donzelot, 1979; Ehrenreich and English, 1979). An unexpected visit from the school attendance officer or social worker raises fears of further judgments on standards of housekeeping and childcare, which may carry very real consequences such as removal of children. Failure to attain the appropriate threshold of knowledge and practice in domestic hygiene or child psychology is heavily sanctioned. On the other hand, competence in presentation of the visible 'home front' attests to general competence in management of the dwelling and its occupants.

THE HOME AS WORKSITE

The home as workplace has been a punchbag for feminists over the years, and this has sometimes obscured the fact that women are nurtured by the home as well as working to provide nurture within it. In writing about women as doing most of the work of running a home, it is not intended to naturalise the gender stereotyping that assigns them this role. It merely reflects the reality of the situation: research has shown that stereotyped gender roles are slow to change even though this might be a 'rational' response to male unemployment or a wife going out to work. One study (Horrell, 1994) showed women not in paid work spending an average of 30 hours a week on household tasks (plus another 18 on childcare) whereas part-time workers spent over 30 hours on household tasks and full time workers over 18 hours. The male partners of these women averaged just under 10 hours a week (with gardening and shopping their main tasks) whatever their wives' employment status. Attempts to fit household behaviour to models of economic rationality have failed dismally, and Horrell (1994, p. 205) concludes that 'the household division of labour would . . . appear to be determined primarily by cultural norms'. Morris (1990) states that 'married women's employment does not prompt a significant rise in domestic involvement on the part of husbands . . . the problem is less one of documenting varied domestic outcomes, than of explaining why change has been so limited' (pp. 189–90).

Even when a couple renegotiates their division of labour, the strategy may founder on the rocks of traditional expectations. It is the woman who can expect a loss in social esteem when the standard falls short. If a child goes to school in a shirt that is dirty and creased it is not the father who is seen by teachers as slovenly. Even if the father is a single parent he is likely to be viewed with sympathy for having neither a woman whose domestic skills he can call on, nor the competence or time to do the task himself. A single mother is seen differently.

The home has been seen as the main site of women's oppression, and the identification of women with the care of the home and its occupants as a factor that limits women's chances of taking their rightful place in public life. Many early feminists especially in America devoted considerable energy to redesigning the home or reorganising domestic labour to free themselves to take a wider role. Some advocated the 'kitchenless house', with efficiently centralised organisation of domestic services. The history of these experiments has been brilliantly chronicled by Dolores Hayden (1981). Few of these solutions involved men doing a share of the domestic labour, as commonly advocated today. Sadly, in the long term they have had little impact. Other feminists have spoken of the tedium and pointlessness of housework (Comer, 1974; French, 1978) and see the home as a prison for women. Certainly the task of running a home, like most work, has aspects which are tedious or sheer drudgery (see Spring-Rice, 1939).

However, this cannot plausibly be seen as the dominant meaning of home for women. For many women who live with others, the work done in the home is seen as an expression of the love in these relationships, so to appear discontented with the burden of housework may be interpreted as falling out of love. The standard of housework is thus a source of guilt and internal conflict for the woman. One can never do enough, but this does not mean that everything a woman does is alienated and oppressive labour. There may be satisfaction at highly developed skills being effectively deployed (Darke, 1994). Hakim (1995) has recently caused controversy by suggesting that women *choose* part-time work in order to carry out what they see as their core role: running a home and caring for their family.

The physical drudgery of running a home has been reduced. Reading Margery Spring-Rice's accounts of the unremitting toil of working-class wives in the 1930s, it is hard to imagine a similar situation today. In a typical case, the wife got up at 6.30 a.m. and was on her feet continuously until early afternoon. Any time sitting down during the afternoon and evening was spent mending, sewing or knitting, and for most of the afternoon and early evening she was again on her feet preparing another meal and putting children to bed. Washday was considerably more onerous.

Surely this has now changed. Take a basic task: dealing with a baby's nappies. Nowadays, fewer women have to deal with so many nappies at any one time, compared to Spring-Rice's multiparous sample. The prewar generation had to wash cloth nappies by hand, with the help of boiling water and soap. In many cases the water had to be fetched from another floor. Mattresses could be protected with rubber sheets but plastic pants did not exist, or later were thought unhealthy, so pram and cot sheets needed frequent washing. In the postwar half-century the task was gradually made easier by the successive introduction of washing machines, detergents, plastic pants, sterilising powder, disposable nappies and finally, all-in-one nappies. Even a generation ago disposables were luxuries for occasional use but now all-in-ones are virtually universal, so that today one may read articles quaintly advocating the use of cloth nappies for environmental reasons and explaining how to put them on, how to wash and where to buy them.

Other tasks have become easier or simply disappeared. Who now darns socks, plucks chickens or turns sheets sides-to-middle? Who, in order to boil water for a morning cup of tea, has to fetch coal and get the fire alight? Much of the physical drudgery has been removed — but the expectations of our knowledge and mental labour are greater. Motherhood does not demand boiling nappies but has become vastly more complex. In comparison with the small collection of rather stern baby books available in mid-century, there are now hundreds, explaining every part of the process in bewildering detail. Although fathers may be spending more time with their children and acquiring some of the necessary expertise in the process, this is seen as a bonus, an optional extra compared with the duty placed on the mother to understand subjects from human nutrition to child psychology. The child, like the home, may be subject to scrutiny by experts.

The work of organising and maintaining the home, too has become elaborated. There are fewer rules, apart from the expectation of cleanliness; there are fashions in styles and techniques for the home interior, and a large number of magazines offering suggestions for imitation. There are rather oppressive expectations of excellence; it is not permissible to be a just-good-enough cook, interior designer or sexual partner. Women are meant to keep track of food that is or is not safe to eat, including eggs, carrots, liver, beef, margarine, whether for the young, the old or the pregnant. Cooking today is expected to involve exotic ingredients, new tools and new techniques and the same is true of sex: failure to come up to the mark may mean dismissal and replacement by a more accomplished performer. Each sphere has its own discourse in which a degree of expertise is called for. Remaining single or living alone may reduce the need for negotiation and compromise but does not eliminate the expectation *on women* for well-developed skills in organising the home. For men there are more options: the ability to engage in the appropriate discourses is one mark of the New Man, but other roles are available: the 'lad', the muddled charmer, and the busy careerist who has no time to develop such skills.

In addition to the greater mental demands the task of running a home now makes, there is now a strong expectation that women should be in paid work as well as maintaining high standards of home management. Whereas in the 1960s and 1970s the home might be seen as a prison or a trap (Gavron, 1968; Oakley, 1974), there is now a greater ambivalence: the home offers delight *and* oppression, privacy *and* isolation, pleasure in nurturing *and* the burden of obligation. The elaboration of household tasks and activities might be seen as a form of 'job enrichment' as advocated by human resource managers. Some may revel in these new opportunities whilst others are exhausted at trying to have and do it all. Others still have no choice: in households where neither partner has a job, and for lone mothers on benefits, the basic tasks are that much harder. The appliances and materials that make housework easier are expensive, to shop efficiently without private transport is difficult, managing on a tiny income requires constant effort.

THE HOME AS HAVEN

A number of writers have discussed the meaning of the home and its contents, including Csikszentmihalyi and Rochberg-Halton, 1981; Rybczynski, 1986; Saunders, 1990; Darke, 1994; Gurney, 1995a, 1995b; Ravetz, 1989; Cooper Marcus, 1995. Differences between individuals in their relationship to the home are patterned in many ways, where gender is one of the most salient dimensions but variables such as age, life history, income level and personality are also significant. Saunders argues *against* the importance of gender but *for* the importance of tenure: only home ownership, he argues, can confer ontological security. Franklin (1990) has presented an opposing argument, describing a couple interviewed at length who consciously rejected the ownership option. Gurney found that men's and women's discourses stressed different aspects of the home. In his sample, men more readily discussed tenure, especially those who were owners (1995a). However, this discourse encompassed negative feelings of home ownership as a burden and a worry as well as its positive meaning as an indicator of success. Almost all Gurney's respondents thought the meaning of home was different for men and women (1995b, p. 12). For women, the home seems to carry meanings that are less about a particular tenure than about links to life events and the emotions that accompany them: marriage, childbirth, raising a family, perhaps bereavement or divorce. Most women choose to share their home with loved ones; it is the container and site of these relationships, and this is its dominant meaning for them.

The home-as-haven may be particularly important for those who face hostility in the public sphere: for example, women members of ethnic minority communities, disabled women or lesbians. They are seen as having even less claim over the public realm than able-bodied, white, heterosexual women, and are thus denied the chance of an authentic selfhood in settings other than the home. bell hooks (1991) has written of the home for African-American families as a nurturing environment and as the site from which an alternative analysis can begin. The very chapter title 'Homeplace: a site of resistance' conveys part of this, and in the course of the chapter she says:

> Black women resisted by making homes where all black people ... could be affirmed in our minds and hearts despite poverty, hardship and deprivation, and where we could restore to ourselves the dignity denied on the outside in the public world. (p. 42)

Lesbians particularly need a haven from the public realm in which they are always on guard against the prejudices of others, including the risk of physical attack. Outside their home or other particular parts of the private realm they are never free to display affection or to talk openly about their own situation (Egerton, 1992). The home is the site where their identity can be affirmed.

For women who are disabled, the situation is complex. In the public realm they are 'handicapped' both by environmental barriers and by public attitudes that deny them their full rights as citizens: to employment, to the use of leisure and transport facilities, to be valued. In public they may either be stared at or find that people look away. In their living arrangements, those with severe disabilities must deal with the assumption by others that they should be in an institution (a Home) rather than living as a householder in a home. (Morris, 1991–2). The need for a home in which the individual is in control is greater in that the public realm is more problematic, and to get this need met may require persistent assertiveness in the face of over-protective attitudes by family and professionals. It is important both that the disabled person is able to get out of the home to use public space and that the home is there as a haven from the constant battles against a maladapted environment and unreconstructed attitudes.

CONCLUSIONS

In all these meanings of the home, there is a complex tension between the home as a source of pleasure and pride, and the home as a problem: too cut off, too much work to manage, never matching the homemaker's aspirations on how it should appear. The mixture of feelings about the home have been brilliantly captured by Clare Cooper Marcus (1995) using the methods of Jungian analysis. Her respondents' conversations with their homes, including the (projected) responses of the home towards the occupier, show how closely the sense of self requires a setting in which we can literally feel at home. Despite (or because of) the complexity of the relationship to the home, the loss of a home can be comparable to the effects of bereavement or marital breakdown. Being homeless profoundly undermines identity and sense of worth. It is to this issue that we now turn.

6.

WOMEN AND HOMELESSNESS

Penny Lidstone

INTRODUCTION

Literature on housing has shown that women are disadvantaged in the housing market, and chapter four outlines the constraints facing women seeking housing. This chapter is concerned with homelessness. It takes as its theme the experience of women applying to the local authority as homeless applicants and later sets this against the assertions by the government that access to housing via the homelessness route is the 'more attractive way' into housing (DOE, 1994a).

The chapter draws upon recent research carried out as part of a postgraduate research degree which focuses upon the experience of making a homelessness application for homeless applicants. It charts the progress of the applicant from making an application to getting permanent housing in order to give the reader some idea of homelessness presentations and of what it is like to 'go through the system'. The responsibilities of the local authority under the homelessness legislation are briefly outlined to show the constraints within which homeless applicants exist, and a discussion of the research methods necessary for this research is followed by the women's experiences of making a homelessness application.

HOMELESSNESS LEGISLATION

The homelessness legislation, introduced in the Housing (Homeless Persons) Act 1977 and now consolidated in the Housing Act 1985 Part III, outlines a local authority's statutory responsibility for homelessness. This is the formal method of rationing housing to the homeless applicant, as the legislation denotes certain categories of applicants who would be eligible for housing. Distinctions are made between 'priority' and 'non-priority' homeless applicants. Priority need is defined as having dependent children or being 'vulnerable'. If an applicant is found to be homeless or threatened with homelessness, in priority need and unintentionally homeless, a local authority has a responsibility for housing. Where an applicant is deemed 'non-priority', lesser duties of advice and assistance are owed. In addition, local authorities are statutorily bound to 'have regard' to the Code of Guidance (DOE, 1991) which 'gives guidance on how authorities might discharge their duties and apply the various statutory criteria in practice' (op.cit., para 1.1).

Government concern that legislation, which should have been a safety net for homeless households, was being used as the 'fast track' into social housing led to the publication of the consultation paper *Access to Local Authority and Housing Association Tenancies* (DOE, 1994a) proposing a review of the homelessness legislation. This document focused attention upon local authority provision for homelessness with its assertion that certain types of households unfairly enjoyed the 'more attractive way' into housing via the homelessness route because 'anyone meeting the necessary criteria . . . will be accepted by the authority as statutorily homeless' (DOE, 1994a, para 2.7). At the time of writing, the government have introduced a housing bill into parliament that will shift local authority responsibility from the provision of permanent housing to the provision of temporary accommodation for homeless households.

THE RESEARCH

During the research the author spent time in a local authority housing department looking at the administration of the local authority responsibility for homelessness. The need to see what was actually happening during the day-to-day administration of the local authority response to homelessness led to the rejection of a survey or questionnaire-based study in favour of a study involving qualitative research methods. Accordingly, semi-structured interviews, participant observation and secondary source materials (such as local statistics, policy and procedure documents) were used to provide the substance and background to the research setting.

The fieldwork centred on time spent in the homelessness section, observing homelessness interviews, going out on visits to homeless applicants and carrying out a series of interviews with housing staff and homeless applicants. The interviews with fifteen homeless women applicants form the basis of this chapter, charting their experiences in making a homelessness application.

Most applicants were in permanent housing at the time of interview although five were undertaken while the applicants were still in temporary accommodation. The applicants had experienced different types of temporary accommodation as they waited for an offer of permanent housing. Some had been in the women's refuge, some in a hostel, some staying with friends or relatives as 'homeless at home', some in bed and breakfast hotels. The circumstances of their homelessness combine with the stark realities of making a homelessness application to give a humbling picture of applicants during times of great difficulty often accompanied by feelings of helplessness and bewilderment.

The interviews may be viewed as providing an illustration of a particular point in these women's lives, and demonstrating the value and importance of the choice of research methods. The use of the qualitative research methods described above gave a depth and texture that could not have been achieved by the use of quantitative methods. I learned the circumstances of the women's homelessness, their fears about their situation, what it was like to make an application, what they felt they could do to affect or improve their housing chances and what it was like for them as they waited in various types of temporary accommodation for an offer of permanent housing.

During the research a process occurred that could be described as the 'evolution of the method'. This comprised a growing awareness of the relevance of feminist research methods as issues surrounding the relationship between researcher and the researched unfolded. I had not started out with a set of research principles that I could have nailed neatly to the mast of feminist research, but as time in the field progressed I realised that my concerns in the field were inextricably bound with issues central to the debate about feminist research 'concerning how to treat participants and how to use the information to which their testimonies give rise' (Maynard and Purvis, 1994, p. 4). I was also surprised (perhaps naively) that there was a debate over issues that seemed central to social interaction and that should be based upon the respect of one individual for another, particularly in circumstances where the researcher is 'allowed' to intrude into the lives of others, often discussing or exploring very sensitive areas. This was very relevant to the areas that I was discussing/exploring as the circumstances of some women's lives were extremely distressing and dealing with homelessness is, by its very definition, dealing with life at the 'sharp end'.

Access to the applicants depended greatly upon my relationship with the housing staff. Had I not chosen qualitative methods traditionally associated with ethnographic studies, it is doubtful whether I would have achieved such a degree of immersion and familiarity in the field or developed a good relationship with housing staff. Consequently, it is arguable whether I would have gained access to the homeless applicants themselves, as this stage of the research depended entirely upon officer consent.

Without the consent of those willing to take part in the research process, much qualitative research would not exist and detailed social analysis could not progress. Indeed, without the consent and co-operation of my respondents, I would not have been able to gain valuable insights into the experience of homelessness. Much research into homelessness centres upon the administration of a system from the view of the administrators; my contribution to this area has been to add another dimension — that of the applicants themselves — a dimension that acquires increasing importance as the government moves to change the emphasis of local authority responsibility for homelessness by shifting the provision from permanent housing to temporary accommodation.

THE INTERVIEWS — WHAT THE WOMEN SAID

The following discussion is based upon the interviews with the homeless applicants. The focus of these interviews was on the experience of making a homelessness application and ways in which the administration of local authority responsibility under the homelessness legislation could be considered as 'informal rationing' (Parker, 1975). The theoretical concepts relating to rationing will not be discussed in this chapter although a fuller discussion may be found elsewhere (Lidstone, 1994).

GOING THROUGH THE SYSTEM

Making an application

One of the clearest pictures that emerged was of the lack of knowledge about the homelessness system or the best way through the 'obstacle course' to gain permanent housing at the other end. The 'prizes' were not always desirable: the condition of the property offered was shocking to several applicants. Some applicants, however, felt extremely fortunate in view of their circumstances to have got a new property at the end of the process.

The applicants had little knowledge of the homelessness route to local authority housing. It was notable that the women had very little information about what to do about their homelessness and none could be described as 'playing the system'. Most applicants interviewed were referred to the council by a variety of means and agencies, such as the Citizens Advice Bureaux, Environmental Health, Department of Social Security, friends, the women's refuge and a building society's solicitor in a repossession case. Two applicants had had previous council tenancies and so had some idea of the local authority role in relation to housing although, again, there was no real understanding of the homelessness function.

I knew nothing, I've never ever been in that position. I mean I've seen documentaries on telly about it, and you know how you hear about people being beaten up in B&B and all that kind of stuff, but I never imagined I'd ever find myself in that position.

I didn't know what was going to happen and I didn't know if I was going to be on the streets or what.

I didn't really know which way to turn, I didn't know *what* to do.

Their lack of knowledge about the system was most apparent through their lack of understanding of the homelessness legislation and the way in which 'priority need' for housing is determined. Some applicants were not even aware that they were being dealt with as homeless or that their case was being dealt with by a homelessness officer. Some 'homeless at home' applicants were not aware that they could be considered homeless and one woman did not realise that her pregnancy gave her priority under the homelessness legislation — she thought that her priority resulted from the fact that she had an address.

When I sat in on the initial application for housing interview, I often heard officers explain the local authority responsibility under the homelessness legislation to the applicants and yet, in my later interviews with the applicants themselves it was clear that they often had no concept of their 'status' within the legislation. The difficulty of a stressful situation in official or bureaucratic surroundings combines to perpetuate a situation of lack of knowledge and misunderstanding for applicants. Thomas and Niner (1989) commented upon the difficulties for applicants of taking in official information, and my work supports these findings.

While lack of knowledge or misunderstanding may indicate a lack of clear communication from officers or that the applicant finds it hard to absorb official 'jargon', it is also an indication that none of these women could be described as 'trying it on' or 'playing the system'. This point also raises questions about assumptions that those 'really in need' (Parker, 1975) and by implication 'genuine' will eventually find their way to the council and be catered for under the homelessness legislation anyway. Lack of knowledge and information can mean that some never find this route — for it cannot be assumed that all eligible homeless households are housed by the local authority — and it can also mean that in many cases the opportunity to prevent homelessness is lost, as people approach the local authority when it is too late to avoid homelessness. This was the case with one woman who relied upon advice and general hearsay from friends. Had she approached the local authority earlier it is possible that the outcome of her housing situation might have been different as the homelessness section could have negotiated with the building society on her behalf.

All the women interviewed were 'really in need' but most did not know what to do and so could be described as finding their way 'by accident' — usually through referral from other agencies. They did not immediately approach the council and for some, particularly in the case of mortgage repossession, such an approach was clearly a last resort. For these women the shame and stigma of repossession was compounded by having to approach the local authority for housing.

The receptionists

All homeless applicants must pass through the reception area in order to gain access to the homelessness staff and it is here that they encounter the receptionists who are the 'external' face of the housing department. The reception area has been identified (Deutscher, 1968; Hall, 1974; Prottas, 1979; Lipsky, 1980; Foster, 1983; Hughes, 1989) as a critical stage in the assessment and outcome of an application for welfare services. Some applicants found the receptionists particularly unhelpful and described them as 'arrogant' and 'unsympathetic'. One applicant likened them to secretaries in doctors' surgeries 'very indifferent to people's problems', and commented upon the obstructive, gatekeeping function that she had experienced.

> I mean I felt like a fish out of water. I felt as if I was a burden to them, you know what I mean? . . . And I kind of felt that I really had to force myself on her to get a result, you know. I think once you get past them, it's not so bad.

She found the receptionists so off-putting that she had wanted to walk out. However she had no other housing options other than to approach the council as her house was being repossessed. Even if unintended, treatment which elicits such responses is unhelpful when applicants and their families are already experiencing uncertainty about their housing futures.

Lack of privacy at reception was commented on by several applicants who said they had found it highly embarrassing having to discuss personal details in a reception area where others could hear what they were saying.

> . . . and it was awful 'cause everyone was stood behind and I was just cringing . . .

> I mean even now you can't help hear what they're saying to people . . . . And you think 'Oh god', you know, 'this is awful'. You know, I mean I actually came out of there in tears the first time I went up.

It was difficult for applicants to be in this situation with receptionists who were perceived as 'more interested in their own work' and who, according to one woman, did not seem to be trained in dealing with people in critical housing circumstances. One applicant commented that the unhelpfulness of the receptionist meant that she and her family were left waiting in the reception area not knowing if she was going to be seen by anyone. She was crying and terrified that they would all be on the streets because the only piece of positive information she had received was that the housing department closed at five o'clock.

The homelessness interview

For several applicants, their experience of the homelessness interview could be described as a type of informal rationing referred to as 'deterrence' (Parker, 1975) and noted by Foster (1983) as an example of the way in which service delivery (in this case the administration of an application under the homelessness legislation) may deter individuals from pursuing a particular service. Given the interview experiences of some applicants, it is remarkable that they continued to pursue their applications and 'survived the system'.

> . . . it was horrible . . . . I came out of there in tears . . . because it was like I was treated like a criminal. [The officer] was like asking these questions, asking that question, but the way [the officer] asked you, it was as though you'd done something wrong.

Several applicants said that they felt that they had to prove they were 'genuine' cases of homelessness and the importance of 'being believed' emerged as a strong theme through the interviews. Applicants spoke of having to prove themselves so that officers knew they were not being dishonest. Some even tried to rationalise such deterrent treatment.

> . . . it just seemed that I was genuine and [the officer] didn't believe me.

> . . . [the officer] already knew about my house on Carfax Road 'cause [housing officer] had told [the officer]. So [the officer] knew I was telling the truth.

> . . . the way [the officer] asked you it was as though you'd done something wrong, but I suppose [the officer's] got to be like that to see if you are a genuine case.

Lack of any other housing alternatives meant that the homelessness interview and assessment procedures were unpleasant experiences that had to be endured in order to get housing.

> Well I didn't like going through it all, and for days after — before [the officer] confirmed they could help me — I was worrying, thinking they were going to ring me up and say, you know, 'You're a liar. We can't help you'. But I had to go through with it because I had to get a roof over our heads . . .

Temporary accommodation

The women experienced different types of temporary accommodation while they were assessed under the legislation or waited for an offer of permanent housing. This seemed to be the most stressful part of the whole procedure for most applicants — a time characterised by uncertainty, difficulties in relationships with friends and relatives and the added stress for those in B&B or a hostel of trying to adapt to an alien culture and environment.

Most applicants were uneasy about the idea of going into temporary accommodation although some were able to avoid it by being considered 'homeless at home' while staying with friends or relatives, including some women who had become homeless through repossession.

> ...if...I'd been in B&B or a hostel, you know, I just wouldn't've been able to cope...I was at a low in my life anyway and that just would've been breaking point.

However, being 'homeless at home' did not mean that waiting for an offer of permanent housing was without difficulty. The women commented on the enormous strain that this put upon their relationships with family, friends or relatives and how uncertainty about the length of time they would have to spend waiting contributed to this.

> ...I mean we were falling out and having rows and, you know, I felt like I had to ask permission for everything and had to keep everything spotless and I had to keep [the child] quiet, you know. It was just really stressful.

> ...while I didn't know [when she would be rehoused] there was the strain, you know, of how long is this going to take and what's going to happen and where am I going to end up?

For the repossession cases, the uncertainty of not knowing when they were getting an offer of permanent housing was intensified by the knowledge that the court had set a date for actual repossession of their homes. This put a corresponding pressure upon them to accept any offer of permanent housing that was made, increasing concern about the type of offer and area in which the offer might be. It is hard to see how a case like this might be described as not in the greatest need for permanent rehousing or to find justification for such cases to be considered for temporary accommodation only, as will be the case if the government's (1996) proposals to change the homelessness legislation goes through parliament.

The worst experiences were described by those in the hostels and bed and breakfast hotels, making it easier to understand Parker's comments that waiting appears to be a 'therapy in itself' and that waiting lists shrink as a result of 'spontaneous remission' (Parker, 1975, p. 207).

Listening to some women recount their experience of temporary accommodation, it was remarkable that they had survived this part of the application process although two women had got 'lost' in the system. One of these women was recorded as having left temporary accommodation voluntarily although this was due to harassment from a third party. Another had refused an offer of permanent housing although it seemed from the interview that she had misunderstood the forms she had to fill in indicating reasons why a certain area might be unacceptable. She had completed it indicating the areas in which she did want to live and was dismayed that she was offered a property in an area that she had not specified. In both instances the local authority discharged its duty under the legislation, leaving both women in a housing 'limbo' with little indication of their future housing prospects. Such cases illustrate how reasons for local authority discharge of duty under the legislation contained in the official statistics may have little bearing upon the applicant's experience and also demonstrate how statistics give no real indication of applicant circumstances.

The loneliness, unhappiness and isolation of being in bed and breakfast accommodation, together with a sense of desperation to get out and of frustration at not being given reassurances of permanent housing in an area near relatives, was cited by one woman as a contributory factor in a suicide attempt. Such feelings of desperation to get out of temporary accommodation and into permanent accommodation were commented on by most of the women who wanted to be settled in their own homes where a more 'normal' lifestyle could occur.

For some women the experience of being in a hostel for homeless households seemed to be like a living nightmare. The picture of temporary accommodation as a culture shock was graphically illustrated during the interviews with women who had experienced this type of temporary accommodation. Time spent waiting in this particular 'queue' was characterised by acute stress, lack of privacy or 'normal' life, and exposure to a different 'culture'. It seemed to be an exercise in survival for the applicants and their families.

For some, being in a hostel was almost like being in a total institution (Goffman, 1968), where the theme of madness and the metaphor of 'prison' was used to describe this experience. Several women found that contact with homelessness staff from the housing department was vital to their survival in temporary accommodation and said that it was important in keeping them sane. The homelessness staff were like prison visitors who brought contact with the outside world, the real world, and for one woman, an almost forgotten world.

> . . . I'd lived there so long in the end, it was like my lifestyle and I couldn't remember what my lifestyle was like anymore . . . In the end you even got

to the point where you didn't want to go outside, just wanted to stay inside all the time.

The thought of 'light at the end of the tunnel' kept her going: the knowledge that one day she and her family would get out of 'prison' was her lifeline to survival. However, this did not prevent her from being very depressed — something also experienced by other women.

The lack of places at the women's refuge meant that some women had to be in other forms of temporary accommodation while they waited to be housed. One applicant commented that a hostel was not an appropriate place for women who were homeless through violence because of the often unacceptable behaviour of other hostel residents. Such women and their children should be somewhere where they could get the help, support and advice they needed — a moot point as the Women's Aid Federation continues to campaign to raise public awareness of domestic violence as an unacceptable feature of our society and of the need for more refuge places.

The picture of temporary accommodation that emerged from the interviews was bleak and intimidating and some women were still affected by reflecting upon this time. One woman felt that her children's behavioural problems had started in the hostel. In this sense, homelessness and being in temporary accommodation particularly, continued to have a profound effect upon family life.

Permanent accommodation

Most women commented upon the constraints of only being made one 'reasonable' offer of permanent housing and indicated that they felt as if they had little choice or control over the procedures. One woman said that being told she would only get one offer and that there was no guarantee that she would not get an area that she did not want, meant that she could not see the point of filling in a form indicating areas in which she did not want to live if there was no real choice anyway. It seemed as if she was being told she had a choice when none actually existed — characteristic of the craziness of her experience of making a homelessness application.

For most applicants the offer of permanent housing could be seen as a highly deterrent part of a rationing process. Not only did they have no real choice, as it was impressed upon them that they would have to accept the first and only offer that they would be made, they usually had very little information about the condition of the property. Consequently, the actual viewing often came as a tremendous shock as properties were often in very poor condition.

It was all boarded up and it looked awful. I just sat and cried. I said, 'I'm not going down there, that's it'. I'm not going down there.

. . . the actual state of the house was just diabolical. Well it was filthy, absolutely filthy. Just looking through the windows it was filthy . . . I mean to give you some idea of how dirty the house was when I moved in, it took my mum *five* hours to clean the toilet.

I hadn't seen it, I didn't know it, and the day I came in I just thought, 'Oh god!'. . . . It was *absolutely filthy*. . . . There was two dogs and two cats lived here and it took me three weeks, three weeks alone just to strip it all down. But I needed to get out of the bed and breakfast before I went demented.

In addition to the condition of the property, some women said that they had found the viewing an off-putting procedure because of the unhelpfulness of the accompanying officer who often reinforced the view that 'someone who is really homeless will accept anything'. One woman recalled how she felt that the officer was 'stand-offish and pushy' with an attitude of 'Well, it's a case of either you want it or you don't' and reinforcing the pressure upon the applicant to accept the offer with 'You do know you won't be given another option?' The applicant's impression of the officer was '. . . all [the officer] wanted to do was show me round and go'. Other applicants remarked that reasonable requests were not met: one woman did not know how to work a central heating boiler, never previously having had central heating. She was referred to instructions written on the boiler when a clear explanation or demonstration would have been preferable. Another woman's partner did not see inside the property until the day the family moved in because a request for another viewing when they could both be present was not accommodated.

The degree of constrained choice seems a highly deterrent element in the rationing process as well as the implied assumption that 'genuine' homeless persons will accept anything and that homeless households should be grateful for anything they are offered. Most applicants felt that they could not refuse 'one offer only' and some said that they accepted the offer as they were desperate to get out of temporary accommodation or, in the case of repossession cases with a repossession date looming, to avoid the possibility of having to go into temporary accommodation. One woman was aggrieved that she and her family had accepted an offer because they had not been informed of the correct date of repossession. She reflected upon whether she might have got a different or better offer without the enormous pressure of having to accept it because at that time she thought they had to be out about a month before the actual date for repossession.

Getting the service?

Although there were times when applicants commented upon procedures during the application process, they clearly had no idea that there were alternatives or that they were not 'getting the service' (Foster, 1983). The issue of 'not getting the service' is closely linked to the rationing concept of 'withholding information' which can result in the dilution of a service. From the interviews with the applicants, and with a working knowledge of the homelessness legislation, it was possible to identify areas where full or better information could have been made available and instances where receipt of this information could have made a difference to some women by reducing the amount of stress and anxiety arising from their uncertain futures. The salient point here is that things were occurring during the application process that, even though the women were not directly aware of them, had direct consequences for the women's experiences of that process.

In one case it was possible that homelessness might have been avoided had advice about repair grants been given. Another woman, homeless through repossession, was worried about the prospect of having to go into bed and breakfast accommodation because she did not know how all her furniture would fit in. Had she known of the local authority responsibility for furniture storage in this event, she might have been less anxious. Two women who found the initial interview distressing commented that seeing a woman officer would have been preferable to a male officer; they were unaware that they could have asked for this facility, and it was not offered.

Stigma and stereotypes

Surviving the homelessness application process and getting permanent housing did not mean that the women emerged unscathed. For some, the experience of, and reflection upon the process was acutely painful because of the stigma that they felt was attached to homelessness.

One woman did not want her children to know that they had become homeless through repossession; as far as they were concerned it was just a case of moving to another house. Another woman, also homeless through repossession, said that she did not even tell close friends what was going on and was grateful that becoming a council tenant was not directly attributable to homelessness so that her new neighbours would not know her circumstances. The reactions of family members demonstrated the stigma of homelessness for some applicants — losing the house through repossession was bad enough, but ending up on a council estate through homelessness was even worse.

Some women commented on the embarrassment of being socially visible as homeless because of living in the hostel. One woman still cringed if she went past the hostel on a bus; she referred to her experience of homelessness as a time of horror and found it difficult to erase the shame and stigma of this memory.

> I think I feel worse now I look back and I've got my own home and my own routine and whatever. I can't believe it. I look back and I can't believe that I did that, that I went to the hostel. . . . I suppose it's a passage in my life but I wouldn't like to tell anyone, 'Oh I was homeless once.' And it's degrading.

In addition to the stigma of homelessness, some women commented that they had felt that others judged them or made assumptions about them because they were homeless and that some homelessness staff made stereotypical judgements about them too. Applicant awareness of the judgmental and stereotypical attitudes often associated with the provision or rationing of welfare services (Foster, 1983; Miller, 1990) was evident during some interviews and one woman explained what it was like to be homeless and a single parent:

> I mean, being sort of in the system as I am, a single mother and stuff, I feel like I'm judged all the time, you know, whatever. Even though now I'm settled in my home and I'm trying to bring my son up the best way I can — and I think I'm quite a sort of sensible person and I've got decent sort of principles and attitudes about life — I still feel I'm judged because I'm bringing my son up on my own. I'm on benefit, you know? I'm in the sort of bracket of, you know, dosser or whatever. So I feel like I'm judged all the time. But when you're homeless, you know — unless it's exceptional circumstances, you know, things like your life's been threatened or, you know, your house has been repossessed through no fault of your own or whatever — people don't want to know, they don't understand. They just think you're incapable, you're not capable of leading a decent life.

CONCLUSION

The experiences of these women form a strong voice that speaks of the disadvantage of being homeless and applying for local authority housing. Their accounts challenge assumptions that applying for housing as homeless is an easy route into social housing and question commonly held stereotypical views that homeless applicants are 'queue-jumpers' and not in the greatest need. None of the women interviewed could be classed in these categories and all would have pursued other housing options had they been available. Indeed, from the women's experiences, it is hard to imagine that this route to housing would be voluntarily chosen or that it could be described as 'the more attractive way' into housing (DOE, 1994, para. 2.7). In the words of one woman, 'I wouldn't wish it on anybody, you know. It's such a horrendous experience, it really is.'

The view of life in temporary accommodation and the bleak and terrible realities of this experience must ring alarm bells for those concerned with the problem of homelessness at a time when the government is poised to overhaul the legislation in the biggest change since the Housing (Homeless Persons) Act 1977. By changing the emphasis of local authority responsibility from the provision of permanent accommodation to the use of temporary accommodation in discharging its responsibilities to homeless applicants, the government is sanctioning deterrence in the provision of housing to homeless households. This will have a significant effect upon the quality of life of those who find themselves without a home and further compound the disadvantage of being homeless. Research concerned with living in temporary accommodation (Conway, 1988; Thomas and Niner, 1989; Shelter, 1995) details the ill-effects of living in this type of accommodation. My own work supports these findings through giving a voice to the homeless. It is time that such accounts and experiences were heeded.

SECTION 3
Challenges for Planning

7.

THE MAN-SHAPED CITY

Jane Darke

INTRODUCTION

This chapter sets out to answer the question of 'how cities work for women.' Women writing on cities have arrived at differing conclusions: for example Elizabeth Wilson (1991) experiences cities as exciting and liberating, an escape from patriarchally imposed identities, whereas others have stressed the dangers for women in cities (Stanko, 1990; Valentine, 1990; Hanmer and Saunders, 1984), or the fact that their physical layout is based on male rather than female use patterns (Boys, 1984a, 1984b; Little, 1994; Roberts, 1991; Lewis and Foord, 1984). This is often attributed to the fact that men predominate amongst urban decision-makers (Greed, 1994).

In this chapter we suggest that cities affect women in three ways, that there are three different discourses around the issue of women in the city. Two of these describe a negative impact on women and one a positive. First we describe the city of property where urban space is seen as belonging to (some) men, and women are seen as a part of men's property, in some cases literally as commodities. Secondly there are accounts of the city as an arrangement of uses in space, an arrangement which may make some types of movement, activities and roles easy and others difficult. This is the city of zones and the way this works against women is analysed. Finally there is the city of diversity, a large, heterogeneous environment (Wirth, 1968) where almost anyone can find others with common interests, where different identities can be tried, where there is always something interesting happening.

Any settlement is an inscription in space of the social relations in the society that built it. Assumptions about roles and the proper place for different categories of people are literally built in to towns and cities, whether those categories are genders, age groups, castes, classes or ethnic groups. Our cities are patriarchy written in stone, brick, glass and concrete. Feminist theories are discussed in more detail in Chapter 1, but it is useful briefly to explain here our use of the term 'patriarchy'. This has been used in many ways (see Walby, 1990; Hearn, 1992) and we do not intend to add to the mass of a discussion which has sometimes prevented the concept from having any analytic utility. Patriarchy is literally the power of the male head of household, seen for example in the old electoral system where the propertied man's vote was deemed to include the interests of his wife, children, servants and any other members of the household (Pateman, 1988). Most writers (including ourselves) use the term figuratively to mean the power of men over women.

Patriarchy takes many forms and alters over time. It co-exists with most economic systems including capitalism, and in many settings: within the household, the workplace, the polity and so on. It is so ingrained in social relations that many people do not recognise it and see male dominance as natural. Individual social actors may resist it and attempt to live their own lives in an anti-patriarchal way, but it affects us all. Various types of analysis can act as a 'discloser' of patriarchy. An analysis of how city space works differently for men and women can reveal that the genders are not different-but-equal but that there is an imbalance of power.

THE CITY OF PROPERTY

Women know that city space does not really belong to them. They know that most cities are dangerous, that they may only use particular parts of the city and at certain times, and that even in those spaces where they are permitted to be (as guests) they must comport themselves in particular ways. Women are excluded from many buildings; in others they may enter but are made to feel unwelcome. There are some men for whom the city is dangerous, who are not fully accepted into the public patriarchy, to whom urban space does not 'belong'. Black men and gay men are at risk of attack by being in the wrong place at the wrong time. Correspondingly, the city is more dangerous for some women than for others. Black women are often subjected to a debilitating barrage of sexist and racist insults. When using the city, women must be constantly on their guard against male appropriation, from the shout of 'cheer up love', the wolf-whistler, the admiring gaze that may give the pleasure of a compliment but is still a form of surrogate possession, to the kerb crawler, the offensive drunk, the flasher, the rapist. As Jos Boys (1984b, p. 84) says, 'women learn that they must *appear sexually attractive* . . . but that they must not *attract men sexually*' (emphasis in original).

Is this form of patriarchy a specifically urban phenomenon? Did the invention of cities create an intensification of male dominance? Was there a pre-urban 'golden age' of matriarchy or at least equality? The archaeological record cannot provide a definitive answer to such questions, and we cannot assume that surviving less-developed societies in isolated regions are similar to the prototypes of our own societies. Nor do animal collectivities, whether beehives or groups of chimpanzees, provide reliable parallels. Connell (1987) and Lefkowitz (1983) warn against myths of a matriarchal origin. However, a plausible scenario will be offered for readers to judge for themselves.

The earliest societies, hunter-gatherers rather than farmers, would have been more-or-less nomadic, but stopping in one place for long enough for the hunters to know where to return with their prey. Connell (1987, p. 145) points out that the assumption that man was the hunter and woman the gatherer is not based on evidence but is a retrospective superimposition of modern gender stereotypes. Nonetheless, among those who did not hunt, young children and the women feeding them would have been prominent. It is possible that this gave rise to gender role expectations which, as today, were only approximately matched by the actual behaviour of men and women. The 'stay-at-home' group would probably have been dominated by women. It is an open question as to which group had most need to develop verbal communication.

This female dominance was almost certainly reinforced in the Neolithic period when plants and animals were domesticated for food. Containment and barriers were now required: to store seed corn, to preserve seasonal food and to protect crops. Mumford (1961, pp. 13–16) equates round, containing forms with the female body (and was criticised for his prurience by Greed, 1994). He also interprets large buildings as phallic symbols, intended to inspire fear and awe. But some of the largest structures that have survived are rounded, earth-covered chambered tombs; the pyramids, also tombs, are somewhat androgynous. These tombs were not occupied as settlements. The dead they were built for, surely important people during their life, included men and women. These structures required a large workforce over decades of construction time, but the social organisation that made this possible is unknown. It must have been hierarchical in that creating a megastructure required direction, but could have been patriarchal, matriarchal or ruled by either/both men and women. Civilisations that did not engage in such large collective projects may or may not have been more egalitarian.

Evidence from pre-urban societies, then, is ambiguous. The development of the first cities may well be related to warfare (Mumford, 1961), where aggression is no longer an individual or small-group act to gain personal advantage, but is an organised activity requiring a chain of command and military leaders, and is aimed to benefit a large group. In regions of frequent wars, cities had to have protective walls and gates, and an internal system of laws and enforcement.

Although the evidence from antiquity is fragmentary, it does seem that virtually all the military and leadership roles were taken by men.

It is plausible that urbanisation coincided with the introduction or at least an intensification of patriarchy. Private property, warfare and paternity between them were sufficient to subjugate women. Property held by others was coveted and fought for. The spoils of war included enemy women; mass rape is still an invariable accompaniment of war, seen recently in Vietnam, Bosnia and Rwanda. The raped woman rarely becomes a possession for longer than the act takes: she may be cast out by both groups. A few are kept by the victors for future sexual use. The regular woman partner has the role of providing legitimate heirs to inherit property and rank. To ensure the paternity of heirs, she must be secluded from other men (Walker, 1983). Women, whether defined as 'good' or 'bad', became part of male property, and were so regarded for millennia. The fact that rape was seen as theft from a husband rather than an unacceptable indignity perpetrated on the woman (and it is still seen in this way in many societies) is evidence for this attitude. Could the male behaviour that many women find offensive today be an echo of this proprietorship?

There is one role in which women have always been present in cities: that of prostitute. Elizabeth Wilson (1991, p. 8) asserts that (all?) women in the city are seen as public women, as prostitutes. Although prostitution may occur in villages, it can only flourish where there is anonymity (or a supportive milieu) for the prostitute and her clients. The fact of prostitution underlines the fact that in the city, women are thought of as property belonging to men. The prostitute is the extreme case of the woman as commodity, although the man cannot possess her feelings even during the short-lived transaction that has been purchased. She acts to resist commodification.

There is evidence for prostitution throughout urban history. The ruins of Pompeii include a brothel, complete with visual aids whereby non-Latin-speaking clients in this port city could indicate their sexual tastes. The social status of prostitutes could be high, as with courtesans or geisha girls, but the virtual impossibility of the courtesan's acceptance into polite society is a recurrent theme in French literature.

Men have always been the dominant gender in cities, as rulers, decision-makers, generals and cultural leaders. A tiny number of powerful women over the millennia are remembered for their unusual qualities: exceptional abilities and determination and as an exception to gender rules (Herrin, 1983; Robins, 1983). But the way male domination was organised has changed over time.

The ancient Greek cities are remembered for their culture and (in Athens' case) as the cradle of democracy. The men participating in the polity were far outnumbered by slaves of both sexes and by women who were excluded from political life. The inspiring civic spaces were the province of men. Private homes were small and mean: the architecture made it clear which people and activities were important. Within the home there was clear demarcation of spaces used by men and those used by women (Walker, 1983).

It is not necessary to trace the course of patriarchy throughout the history of urban settlements. There is no time when the 'public realm' as defined at the time (the church, the court, civil society and so on) was not dominated by men. This dominance clearly applied in the households of wealthy and powerful men. There was less reason why patriarchal relations should prevail in households without property. Where a household's survival or standard of living depended on selling one's labour, the contribution made by women would have been important and might have conferred almost equal status. Yet until recently women were excluded from political participation. Most jobs were closed to them and particularly those with reasonable pay and status (see Chapter 2).

Even now that legal barriers to women's access to jobs and particular buildings have (largely) been removed, women are excluded from urban space in ways that may be more difficult to negotiate than the explicit barriers. Most women have experienced the not-so-subtle behavioural signals that tell them they are trespassing on the territory of men. These signals include the jeering, catcalls and explicit sexual insults that greet women MPs who speak, the various stratagems employed by male workers in a print shop (Cockburn, 1983), pin-ups, horseplay, bullying or sexual harassment in other worksites. Professional bodies reluctantly admitted women but did little to make them feel welcome (see Greed, 1991, and Chapter 12 of this volume). It required energy and organisation to gain the most basic acknowledgement of women's presence: Greed (1991, p. 79) describes how the Royal Institute of Chartered Surveyors only provided female toilets as late as 1973. Disabled people are still waiting for this recognition of their existence in many public buildings. Women may simply be ignored: they may be in the workplace in senior posts but are excluded from an inner circle of advisers and confidants. A lone woman in a pub, hotel or restaurant provokes unease in waiters and other staff; she herself may be unable to relax due to the risk of unwanted attentions or remarks. The urban environment constantly gives women messages about their place.

Messages come not only from men who are present, even if only as passers-by in the street or readers in a library. The urban environment is bursting with visual images on hoardings and in shop windows. Visual and behavioural messages about women's 'proper' role are reinforced in the privacy of our own home by television, especially the advertisements. Even the newly-knowing ads that subvert stereotypic expectations depend on our recognising the stereotypes that are being both undermined and reinforced. Women out of place are a strong subtext or even the main storyline in many films. Visual images reinforce the commodification of women, which is seen not only in prostitution but also in other roles in the sex industry as well as the 'trophy wife' whose wardrobe, grooming and 'emphasised femininity' (Connell, 1987) serve as confirmation of the husband's status and wealth.

Just as women were the property of men in ancient Athens, men in cities still feel they have the right to take possession of women, although women may resist this appropriation. Although there are some senses in which there has been a move from private to public patriarchy (Walby, 1990), it is also true that both public and private patriarchy have been present throughout urban history. Private patriarchy is far from superseded: at worst it is being exercised as domestic violence (Dobash and Dobash, 1980). Male propensity to violence appears to be associated with exaggerated stereotypes regarding gender roles (Scully, 1990), just as does behaviour in the public realm that shows women that men still regard urban space as their own, and women as a type of 'non-conforming use'. This is what we mean by the city of property.

THE CITY OF ZONES

The city of zones is a more straightforward concept. Most cities and especially modern, planned cities, have distinct areas within them: residential suburbs, industrial estates, shopping areas, office areas and so on. These are linked by public and private transport routes. Various authors, mostly women, have analysed how zoning patterns are based on stereotyped gender roles, where households have a male breadwinner who works conventional hours, and a female housewife who uses the urban environment in a different way, taking children to school, shopping, spending most of her time caring for the home and other household members (Tivers, 1988; Pickup, 1988; Lewis and Foord, 1984, Hanson and Pratt, 1995).

Cities have always been spatially differentiated but the city of zones developed particularly during the nineteenth century. At this time, the industrial revolution and the start of rapid urbanisation were radically changing British society. In the course of the nineteenth century, gender roles became much more differentiated and at the same time (and of course no coincidence), cities became more complex spatially, with separation of functions and the start of large-scale suburbanisation.

Davidoff and Hall (1987) show brilliantly through case studies of particular families how roles and locations for middle-class families changed over this time. Richard and Elizabeth Cadbury started a draper's business in Birmingham in the 1790s, and lived over the shop. Together they ran the business; Elizabeth also organised the large household (there were ten children as well as apprentices and servants) and helped to maintain the kinship network that directly supported the business through investment and professional advice. They established a second home on the outskirts of the city in 1812 but did not move out of the house above the shop until retirement. Their newly-wed son moved in to the shop, but moved to the suburbs much earlier in his married life than did his parents. His wife's role in the business was less active but she had to give emotional and moral support to her husband, children and the wider kinship network. The following generation, in the late nineteenth century, showed the bifurcation of gender roles even more strongly. The then manager, George Cadbury, would not allow married women to work at Bournville. Thus middle-class precepts of family organisation were transmitted to the working class. There is mixed evidence on whether these changes were welcomed or resented.

These twin separations, of gender roles and of homes from workplaces, were taking place all over urban Britain. Men's occupational roles were becoming both more specialised and a more important component of identity. In the public sphere, politics and government could no longer be left to gentlemanly amateurs but became a full time occupation, supported by a bureaucratised civil service. Here was the rise of public man. New professions appeared alongside new ways of regulating and organising traditional professions. The expansion of manufacturing was accompanied by the rise of specialist financial services. The increasing diversification of society was reflected in the proliferation of building types, with factories, offices, railway stations, clubs, professional institutes, universities, stock exchanges, concert halls, town halls, sports grounds, libraries, shops and department stores.

Only the last three of these semi-public buildings were available to women on something like the same terms as men, as spaces where they belonged (see also Chapter 5). Suburbanisation was the physical expression of what was seen as women's main role: as wife and mother to create a haven from the bustle of public life. Even women who were not in fact wives and mothers could be called upon by a male relative to 'keep house' for him. Older women could call on the support of younger female relatives. The expectation that their role was to create a clean, sweet-smelling and moral centre was applied, as the century progressed, to working-class women as well as to the middle classes: see the writings of campaigners such as Lord Shaftesbury, John Ruskin and Octavia Hill. For example, Shaftesbury fought against women's employment, saying that 'if you corrupt the women you poison the waters of life at the very fountain'. Octavia Hill, despite her own work managing housing, was opposed to women's suffrage, believing that 'men and women . . . have different gifts and different spheres' (Hill, 1910).

Until late in the nineteenth century, the pattern in which urban uses were laid down in space was little influenced by planners: market forces (themselves patriarchal, as men were the property owners) were moderated only by building regulations and bylaws. The town planning movement had its roots in disquiet at the chaos and urban problems associated with the rapid urban growth of the industrial revolution, although people had been drawing idealised city plans for long before that. The overcrowded and insanitary conditions of the working class was what ostensibly concerned the movement. Similar conditions had distressed only a few philanthropists when they had occurred in rural settings in earlier centuries. When they occurred in cities their effects went beyond the household concerned; they could impact directly on the whole population. Poor sanitation could spread infectious and fatal diseases; faulty socialisation into proper sexual standards could corrupt other classes, and there was the danger of crime and urban unrest. When poverty was thought to result from a moral failing of the poor, any intervention was considered counter-productive, but it gradually became clear that many were poor through no fault of their own (Rowntree, 1901). This allowed for urban interventions to ameliorate living conditions, initially through clearance of insanitary dwellings or areas and from 1890 by building homes to replace those cleared. From that time the scope of planning powers increased in most countries, at least until the rise of monetarism and the concomitant anti-statism.

Town planning has not had a single monolithic value system but there have been some hegemonic assumptions within the profession. Not surprisingly, the professionals tend to accept the prevailing social assumptions of their day, but if these are then built in to urban form, their effects may persist after social patterns have changed. Planners have tended to share the dominant perception of the late nineteenth century that saw women only as wives and mothers. In one sense it was progressive to treat this role as work and to act to improve working conditions. Being moved from a slum to a setting offering more light, clean air, a water supply and self-containment for the family would indeed represent an improvement for working-class wives (Spring-Rice, 1939). However, they might also have missed the corner shop, the emotional and practical support of nearby kin and neighbours of long acquaintance (Young and Willmott, 1957) and the variety of opportunities for earning to supplement the insufficient wages of some men. For single women, the disadvantages could well have outweighed the benefits, but they were rarely catered for in planned environments (there were some exceptions such as Homesgarth at Letchworth: see McFarlane, 1984).

Marion Roberts (1991) has shown how strongly the post-war planning of London was influenced by fears about the low birth-rate in the interwar years. Women were to be encouraged to reproduce through re-housing in medium density houses rather than high density flats. Policy-makers' concerns lest women should succumb to the rival attraction of other lifestyles such as an equal role in the labour market are seen in the Beveridge report (1942) which spoke about the continuation of the British race, and in the contemporary writings of Mumford (1945).

This mixture of gains for women in their conventional role, together with a failure to cater for any other role, has marked planners' impact on city form. The post-war new towns, seen as some of planning's finest achievements, prioritised male employment. The neat separation of different spheres of life allowed for conventional roles but not for the combination of wife, mother and worker, trying to buy household necessities rather than relaxing in the works canteen in the lunch hour. Where women wanted to work, perhaps part-time, this was difficult because of poor off-peak public transport to the industrial estates and the lack of shopping facilities nearby (Lewis and Foord, 1984). Residential areas in new towns and overspill estates were premised on the male breadwinner and non-employed wife: lack of childcare and local kin networks and the long and expensive journey to work made other roles almost impossible.

Horizons were limited for the women left all day in these suburbs, now experienced too by men who have lost jobs in the recession. Suburbia may be an agreeable haven for the working man but a nightmare for women or men seeking work, with restricted opportunities for employment, leisure and shopping in the vicinity and no means to travel to better opportunities. It may suit young children but can be intensely boring for the adolescent without personal transport and with limited means. It can be stifling for anyone whose face doesn't fit. Only recently have planners attempted to bring residential uses back into city centres, offering more roles and opportunities for women. We return to this issue in the next section.

Gender stereotypes held by planners are no more extreme than those current in general discourse. Women who find that cities fail to meet their needs should blame patriarchy rather than male planners. Planners are not aware that their land-use maps arise not only from a rational decision-making process but also from social stereotypes. We cannot call this the 'city of rationality', an initially attractive title, in view of its limited perspective on women in cities. The city of zones compartmentalises activities such as work, leisure, travel and home life, which most women do not separate out in this way.

THE CITY OF DIVERSITY

Despite the problems for women that the city of property and city of zones present, the city is the place where many want to be. The city is where opportunities are, not only for work and careers but also to become a more interesting person, to try different roles outside the confining labels of a face-to-face society. Cities offer choices: choices between anonymity and affiliation, choices of entertainment, job choices, identity choices. These choices are particularly important for women because of the limited range of roles they are offered in other settings. Some feminists have spoken of 'compulsory heterosexuality' and of the pressures on women who choose not to have children, and there are also strong norms about the presentation of self and home. The city offers an escape from stifling expectations. It can be lonely but is also liberating, especially for young women making a break from the prescriptions and presumptions of the family.

The need for any young woman with ideas of her own to make such a break is a strong underlying theme of the novels of Charlotte Brontë and George Eliot, and was later taken up by male novelists such as Gissing, Wells and Lawrence. Modern women novelists virtually take the city as backdrop for granted as a place where a central female figure can be or becomes self-determining, with like-minded women friends as indispensable support and undependable men in walk-on roles (Lessing, 1973; French, 1978, and too many younger writers to cite individually). The best books on the pleasures of cities are by women (Jacobs, 1961; Wilson, 1991).

In the city, we can act with others to challenge the status quo. Wilson has written of the city as a perpetual carnival: this is the atmosphere on a street demonstration, even though the causes are serious, like the campaign for the vote. The suffrage movement was a typically urban phenomenon: a mass campaign that involved appropriating city space in new ways, being in what the patriarchy would see as the wrong place at the wrong time, chained to railings, out at night breaking shop windows, shouting outside Parliament or politicians' homes, interrupting men at meetings. The very act of coming together to organise the suffrage campaign, the unaccustomedness of being in a public meeting or on a march with hundreds of other women, must have generated an extraordinary elation. This exhilaration can still be powerfully felt when groups that cannot freely use city space in 'normal' circumstances come together in carnival atmosphere for particular events such as 'reclaim the night' demonstrations, lesbian and gay pride marches (Valentine, 1995 p. 108; Connell, 1987, p. 133) or the newly militant groups of disabled campaigners holding up the traffic.

The city has always been important to ethnic groups that are minorities in a larger community: only in the urban setting is there a sufficient social group to allow valued cultural identities to be maintained. These identities can also be changed, challenged and alternated: Hanif Kureshi's *The Buddha of Suburbia* (1991) shows a male central figure and his father doing just this, the very title a comment on urban roles and settings. There are parallel possibilities for young women. The city may be a positive choice for minorities or it may be the only place where a sufficient size of group exists to permit a social and cultural life. At the extreme there is no choice, as in the ghetto. Just as the city is patriarchy in built form, its social areas are also a map of a racist society (Smith, 1989). Whilst the city offers diversity there have been many victims of those who use force to reduce diversity and to reassert their own dominance.

For lesbians and gay men, only in the city, among a significant peer group, can these identities be affirmed or celebrated rather than problematised (see Wilson, 1991; Valentine, 1995). Valentine talks of the diversity within a lesbian 'community' which does not constitute a face-to-face group, rather a loose network which individuals can use to locate compatible friends. The space that is claimed back from hegemonic heterosexuality is partly real, partly symbolic. Although lesbian households constitute a minority in the area Valentine describes, 'the diversity of lifestyles . . . with significant populations of students and ethnic minorities . . . appears to allow lesbian households to remain more anonymous than they would be in . . . predominantly white nuclear family-oriented middle class estates' (p. 98). The lesbian network is 'largely invisible to those who aren't in the know' (p. 100) but support groups are particularly important for lesbians and for any other groups who may feel isolated by the prejudices of others. Clearly the city provides potentialities for affirmation and affiliation for minorities especially those who are stigmatised and widely misunderstood, but may be equally important for women who merely choose to live without a partner.

The diversity of cities is what provides choice and excitement to all these groups, but it is threatened. The increasing polarisation of income levels and life chances (see Chapter 3) not only in Britain but in much of the world including the countries where communism has collapsed, carries the threat of instability. Social polarisation has affected the quality of life in many inner-city areas. An area that is vibrant when there is a social mix becomes difficult to live in when most residents are poor and unemployed. Crime rises, ethnic diversity turns into racial tension, those who can will move out. Women do not feel secure even in their homes. Research findings based on fieldwork in the 1970s, showing that women prefer the inner city to suburbs (Saegert and Winkel, 1980; see also Fava, 1980) may no longer be valid. Some inner-city residents are able to gain security through guards and barriers, but we would all prefer to live in a society where these were not necessary. The city of (men's) property with extreme social divisions between the haves and the have-nots is a threat to the city of diversity. Social divisions, when reinforced in spatial separation, pose the danger that others are not seen as in any way like ourselves, not acknowledged as having any shared membership of a society. Misrepresentations based on ignorance about single mothers, council tenants, lesbians or people of Asian descent are a part of this. The stigmatised and disadvantaged groups, concentrated in a neighbourhood that others dare not enter, are seen as a threat to social order. The challenge of urban life is to find a way of valuing diversity and its spatial expression in a way that is inclusive rather than exclusionary.

8.

WOMEN'S SAFETY

Helen Morrell

INTRODUCTION

This chapter explores the concept of women's safety. Feminist theories of male violence and the gendered structure of crime are reviewed and considered in the context of urban problems and policy-making. The chapter then goes on to look at different approaches to women's safety, drawing from examples of current practice. This exploration of women's safety is guided by a feminist understanding of violence against women and by tacit recognition that, although great steps have been taken in providing safety for women, from the first steps taken by feminist organisations such as Women's Aid and the women's refuge movement to more recent approaches characterised by local multi-agency partnership working, there is still a long way to go before women's safety is tackled at the level it necessitates and deserves.

DEFINING WOMEN'S SAFETY

The term 'women's safety' holds much potential but is commonly informed by a misguided view of women's experience of and fear of crime so that the strategies which draw on it may have limited success. Common interpretations of women's safety, which are often reinforced by both the media and in policy development, place an emphasis on 'stranger danger'. This interpretation is based on an assumption that most attacks on women take place in public places and are committed by someone who is not known to the woman. This 'stranger' is often conceived of as someone who has little or no control over his actions and the attack is therefore seen as both unpremeditated and opportunistic.

This stereotype is doubly problematic. As Kelly (1988) observed, on the one hand it serves to deny any violence that doesn't fit the stereotype, such as domestic violence, and, on the other, it pathologises the male offender 'resulting in both a deflection of responsibility from men and the denial of women's experience', (p. 36). The limitations of this stereotypical definition of violence against women mean that a great deal of violence goes unrecognised both officially and culturally. In the past decade, however, radical feminists such as Kelly (and also see Hanmer and Saunders, 1984; Stanko, 1985, 1987, 1990, 1993; and Radford, 1987) have been instrumental, through extensive empirical work, in problematising this definition and in unravelling the true nature of violence against women.

The radical feminist perspective draws attention to the fact that a great deal of violence against women is 'hidden' both in terms of women's reluctance to report it and in terms of the traditional crime survey's inability properly to account for it as it is really experienced by women. 'Traditional' crime surveys include those undertaken by the Home Office to derive 'snapshot' pictures of the national picture of crime and used to inform national policy on crime prevention, law and punishment. The British Crime Survey is carried out at intervals of around three years and uses a large-scale quantitative survey to assess rates of reported crime. The approach to criminology which has developed around this survey instrument has been labelled 'new administrativism' by feminists and by New Realists (see Young, 1988 and 1992). The New Administrative criminologists are criticised by these groups for, amongst others things, generally underestimating the problem of crime.

From a feminist point of view, the British Crime Survey does not properly account for violence against women because it underestimates domestic violence and sexual offences. Such crimes are notoriously under-reported and the nature of the survey does not provide scope for women to record the 'everyday experiences' (Stanko, 1993) that are outlined further on in this chapter and which contribute to a general climate of fear amongst women. Consequently the conclusion which has repeatedly been drawn from a succession of surveys — that women's fear is far greater than the real threat of violence against them as borne out in the crime statistics, and in this sense irrational — is challenged.

A number of feminist surveys have been carried out with the express intention of discovering a more accurate picture of crime against women. They include Jalna Hanmer and Sheila Saunder's study of West Yorkshire (1984); the *Wandsworth Violence Against Women — Women Speak Out Survey*, co-ordinated by Jill Radford (1987); and *The North London Domestic Violence Survey* (Mooney, 1993 and 1995). These surveys use a broader definition of male violence and one which is informed by women's (self-reported) experience. The North London survey, for instance, defines sexual harassment as 'being kerb crawled, followed, approached or spoken to and shouted after in a way that was perceived by the respondent to be sexually motivated' (Mooney, 1995: p. 42). A high proportion of women interviewed (42 per cent) reported this.

The findings from this survey overturn the commonly held myths that women are safer in their own homes than in the public sphere and are at most risk from strange men. These myths were in fact true for men but for women the exact opposite of the findings held true:

> For a man the public sphere is twice as likely to be the arena of risk in comparison to the home; for woman the pattern is exactly the opposite ... for a man, strangers are the greatest risk, followed by acquaintances and then partners: the risk decreases with intimacy. For a woman, partners are by far the greatest perpetrators of violence ... the reverse is therefore true'
> (Mooney, 1995: 49)

More qualitative feminist research has attempted to 'get beneath' the crime statistics and to uncover the nature of women's fear of violence and the role it plays in their everyday life. Kelly (1988) and, elsewhere, Stanko (1985) talk about a 'continuum' of violence against women which ranges from intimidation and threats of violence to sexual harassment and rape. Women's experience of violence is not a constant along this continuum but will depend on their particular circumstances and characteristics. Women with disabilities and women from non-white ethnic backgrounds in particular may experience a range of different threats within this continuum.

This continuum is useful in documenting and acknowledging the 'many incidences women experience as abusive (that) are not defined legally as crimes' (Kelly, 1988, p. 77). Stanko's work in particular has been valuable in both acknowledging and trying to define what she terms the 'everyday experiences' which contribute to women's fears of sexual threat and intimidation. Such experiences, which are often trivialised, include flashing, peeping toms or incidences like strangers in the street making lewd comments as they walk past. The range and prevalence of male behaviour which is seen by women as threatening and which contributes to a climate of fear amongst women, leads Stanko to conclude that 'women's everyday lives are permeated by intrusions by men' (Stanko, 1993, p. 157). New Realists too have acknowledged the alarming extent to which women's lives are intruded upon in this way. Jock Young uses the following analogy to make this point: 'The equivalent of sexual harassment for men would be if every time they walked out of doors they were met with catcalls asking if they would like a fight' (Young, 1992, quoted in Mooney, 1995, p. 6).

Few would challenge the assertion that women's fear of crime and their concerns about personal safety are unfounded when the extent to which they limit their participation in everyday life is considered. Most women rarely go out alone after dark and, if they do, are restricted in the type of public spaces they can inhabit. Both during the day and after dark women adopt a variety of precautionary measures which range from complete avoidance of leaving their home at night to carrying weapons and changing their travel patterns when undertaking regular journeys.

Stanko's research, outlined above, and Kelly's research into sexual violence (1988), provide interesting accounts of women's interpretations of their experience of male violence. Green et al.'s study of women and leisure activity (1987), which is discussed at greater length elsewhere in this volume gives a shocking account of the extent to which women's lives are constrained by men's control over their use of public spaces. The women in this study reported a wide range of control over their participation in non-work and non-domestic activities, both by their partners and by men with whom they came into contact when in public. The women reported 'a range of behaviour used by men, from silent disapproval through a variety of joking and ridiculing behaviour and sexual innuendo to open hostility' (Green et al., 1987b, p. 88) as ways in which they were discouraged from participating in social life.

In all the studies listed above the concept of social control is central to the analysis of violence against women. The various intimidatory strategies outlined are recognised by both the researchers and the researched as means by which men attempt to exert control over women on a day-to-day basis. It has been argued that, in this context, the reluctance of the formal criminal justice system to intervene in the area of violence against women can be seen as the state's endorsement of this more informal routine policing by men (Radford, 1987). The outcome of this, from a radical feminist point of view, is a vicious circle in which women are forced to seek protection from those men who, in some cases, may represent the most real and greatest threat to them. This paradox, which is at the core of the women's safety debate, may mean that any strategies which do not attempt to address it will, as I have posited in the introduction to this chapter, be ultimately doomed to failure. The existence of the paradox, as well as presenting limitations, presents a challenge in terms of finding appropriate women's safety interventions which are based on more realistic understanding of violence against women and the nature of women's fear of crime.

THE DEVELOPMENT OF WOMEN'S SAFETY STRATEGIES

The development of women's safety strategies has paralleled the growth in crime prevention activity which has taken place in the last fifteen years. The key characteristic in this has been a drive from central government to move the responsibility of crime prevention activity outwards from the police and the formal criminal justice system to other agencies of social control and to the public at large. The 1984 Joint Circular on Crime Prevention (Home Office, 1984) marks the beginning of this movement to promote partnerships against crime and to place crime on a broader public policy agenda. The 1989 circular *Tackling Crime* (Home Office, 1989) expanded on the idea of a partnership approach to crime which culminated in the Morgan report *Safer Communities: the Local Delivery of Crime Prevention through a Partnership Approach* published by the Home Office in 1991. This report represented a watershed in terms of attitudes towards crime prevention activity and recommended, amongst other things, a movement outwards from the situational 'locks and bolts' attitude and advocated experimentation with more social approaches which would emphasise the role of community ethos and education:

> It is the view of the Working Group that the social aspects of crime prevention, which seek to reduce those influences which lead to offending behaviour, and the fear of crime, need to receive attention at least equal to that given to the situational aspects of crime prevention, in which efforts are made to reduce the opportunities and 'harden' the potential targets for crime (Morgan, 1992, p. 13).

This represents a broadening and deepening of the concept of crime prevention in which new agencies and new types of interventions are brought on to the crime prevention agenda. Most cities have risen to this challenge. In some cities the Home Office's Safer Cities initiative[1] has provided the impetus for this type of approach. In other cities local authorities have taken a lead role and established their own multi-agency safety partnerships.

Such safety strategies, which focus on individual communities, often incorporate policies directed at particular population groups, depending on the perceived needs of the community. Many community safety strategies have policies on women's safety. These vary considerably from the inclusion of stand alone interventions to comprehensive women's safety strategies which comprise a number of co-ordinated interventions. The Derby Safer Cities women's safety strategy, now led by a community safety partnership, is a good example of the latter, incorporating a number of different interventions including safe transport, community-based support for women in localities severely affected by crime and support for survivors of domestic violence. The interventions are co-ordinated by a number of committees comprised of representatives from the agencies involved.

The London Borough of Hammersmith and Fulham, also part of the Safer Cities scheme, has an exemplary strategy focused on, although not restricted to, survivors of domestic violence. This is part of a corporate community safety strategy which cuts across the work of a number of local authority departments and partnership agencies. It includes an educational and publicity element and a number of projects which provide a range of support, both crisis intervention and longer term, to survivors of domestic violence. Both strategies outlined adopt a holistic approach towards women's safety, providing a combination of social and situational interventions and catering for both the short-term and the longer term needs of women experiencing violence. Both strategies also incorporate programmes which work with male perpetrators of violence. A small number of such schemes exist nation-wide.

CHANGING MEN

Radical feminists argue that little real gain can be made in women's safety unless men's behaviour is challenged. Liberal feminists would also argue that educating men about issues of violence plays a crucial role in preventing violence against women. A number of theories have been posited to explain male violence and there is not room to discuss them here. For a succinct summary see Scourfield (1995) and for a critical overview see Stanko (1994). Broadly, categories which have been recognised include biological explanations, which claim that men's violence is natural; systems theory, which shifts the emphasis on to a social level and sees certain couples or family units as 'violence prone'; and social learning theory which claims that violence is 'learnt' from observation of others. Violence is also 'learnt' in the sense that it is found to achieve certain desired ends such as ending an argument or gaining power. The feminist explanation of violence places gender conflict at the centre of the explanation and sees men's violence as a means through which male power over women is maintained.

Approaches to working with men differ according to the model of violence used to inform the scheme. Schemes directed by a feminist understanding seek to increase men's awareness of their use of power within relationships and to explore their understanding of masculinity. Scourfield draws attention to what Clatterbaugh (1990) calls radical and liberal pro-feminist approaches. The latter would see men as damaged to some extent and thus as victims of their own behaviour and of their masculinity. Although not adopting a liberal approach, Stanko (1994) explores the need to focus on men's 'individual' violence and the nature of masculinity in order to understand why men are violent.

Strategies which aim to change violent men are usually seen as part of a broader women's safety strategy and linked to agencies providing support for women. It can be argued that support work with women indirectly challenges and re-educates men by encouraging them to be less tolerant of unacceptable behaviour. This stance has always been problematic from a feminist point of view because it has at its core notions of victim precipitation: that certain women are prone to form relationships with violent men. It also suggests that changing female behaviour, by encouraging women to become more assertive and less dependent, will ultimately change male behaviour, which clearly is not the case. Nevertheless, schemes which offer women education and information on domestic violence can play an important role in women's safety.

COMMUNITY-BASED SUPPORT

The author's own research, conducted with Sue Yeandle, into a Safer Cities women's safety strategy, included an in-depth study of an intervention aimed at providing education and long-term support to women. This women's support centre was based in a high crime community and one which was identified as being excluded from a range of statutory services. The centre, run by the voluntary sector and funded by Safer Cities and other partnership agencies, provided a drop-in centre and crèche run by local women for local women. A range of activities were organised on a day-to-day basis alongside access to support and counselling. The aim of the intervention was to provide a mechanism through which networks amongst women in the community could be strengthened, to provide mutual support and, in the longer term, empowerment of women.

The voluntary sector has an important role to play in filling the gaps left by statutory agencies such as the probation service and social services; most notably those of longer term support. The voluntary sector may be also best placed to deliver interventions which are community based, women-defined and responsive to the particular needs of women in the community. Such interventions will also provide a link between women and the more formal support services which they might otherwise find difficult to access. The particular women's centre discussed here played a crucial role in providing women with a safe place to consider their options and, through links with local refuges and other support systems, provided them with both the information and the emotional and practical support they needed to take their own decisions about their own future.

Such social interventions are innovatory. Because of the nature of their objectives, which are rooted in personal and community development, they will necessarily take a long time to provide tangible outcomes in terms of reducing crime. Any evaluation, then, must be set in this long-term context. One consequence of the long timescale involved in the measurement of outcomes of such interventions may mean that, in the competition for scarce crime-prevention funding, shorter term 'nuts and bolts' schemes are favoured. The challenge of convincing agencies involved in crime prevention of the benefits of social measures in women's safety remains.

PLANNING AND WOMEN'S SAFETY

Planning and design issues play a critical role in women's safety. The relationship between crime and the built environment has long been the subject of debate. Newman's defensible space theory (Newman, 1973) and Alice Coleman's analysis of urban form (Coleman, 1985) have both been influential. These works explore the role of public and private space, of building layout, land use and other design issues in contributing to both the fear of crime and to crime itself. Several locally-based surveys undertaken during the 1980s confirmed that fear of crime was far more prevalent among women than men.[2] Fear of crime can severely inhibit and restrict women's freedom of movement. Valentine's work (1990, 1992) explores the relationships between: 'women's fear of male violence and their perception and use of space'. She found in her survey work in Reading that women produced mental maps of feared environments and dangerous places which inhibited their use of space.

Environments where women have reported that they feel most at risk include: deserted bus stops, railway stations and footpaths; pedestrian subways; multi-storey car parks; blind corners; lifts; stairwells; and alleyways (see Atkins, 1989; Valentine, 1990; 1992; Crime Concern, 1993; Women's Design Service, n.d.; Trench et al, 1992). Factors which increase fear in these places include: lack of activity, surveillance and visibility; poor environmental quality, design and maintenance; and the existence of grafitti and violence (Atkins, 1989).

Some of these ideas have influenced and been drawn together in the Department of the Environment Circular *Planning Out Crime* issued in 1994 to provide guidance to local planning authorities in development control and in generally promoting a safer built environment. Most importantly, it established crime prevention as a 'material consideration' in the assessment of planning applications. It also stressed that crime prevention should be 'one of the social considerations to which regard must be given in development plans (DoE, 1994b, p. 3). Several local authorities have already begun to take this issue seriously. Authorities such as Leicester, Greenwich, Lambeth, Manchester and Southampton have produced design guidance and advice covering such matters as layout, footpaths, housing areas, open spaces, landscaping, lighting and planting.

Although playing a critical role, planning as an activity remains severely restricted in the extent to which it can intervene in safety issues. Design solutions are based on a situational model of crime prevention which is guided by an understanding of crime as opportunistic. Its success rests on 'target hardening', or eliminating opportunities for crime, and in this sense is limited in terms of both the type of crime it can effectively tackle and the types of communities and individuals it can effectively serve.

The 'siege mentality' which is at the heart of this approach — that communities, with appropriate target hardening, can protect themselves from external threats — poses problems when applied to high-crime communities, which are invariably characterised by both high rates of offending and of victimisation. In these instances the threat of crime is from within the community. Here more social approaches to crime prevention are necessary to attack the source of criminal activity. For all types of community, the approach is also powerless in preventing crimes such as domestic violence where the threat is from within the home. This is recognised in the following statement:

> Clearly the built environment alone will not stop those who are determined on criminal activity, nor will it reduce crimes like violence in the home or fraud which are not related to the environment (Trench et al., 1992, p. 281).

Notwithstanding this, the built environment plays a paramount role in women's perception of risk. Jane Darke, in Chapter 8, considers the relationship between women and their environment and outlines the extent to which the built environment has been planned and shaped with men's needs in mind, and the consequences of this for women.

SAFE TRANSPORT

The neglect of women's needs has meant that their use of the city, both at night time and during the day, is severely curtailed. Rosalie Hill's chapter outlines the pivotal role that transport can play in this and describes the ways in which the particular needs of women have been neglected in the planning of both public and private transport. There have been a number of attempts to address this which have included encouraging co-operative working between transport authorities, companies and passenger organisations to promote women friendly services.

Dedicated women-only transport provides another option and has been pursued by, amongst others, Bradford City Council in partnership with Bradford Safer Cities, Coventry City Council and Derby Safer Cities. The latter scheme, a women-only taxi company, was the subject of an evaluation carried out by the author and Susan Yeandle over a six-month period. The evaluation showed a high demand for the service amongst local women. A survey of users concluded that the scheme was highly valued and enabled many to make journeys which they would not otherwise have been able to make. The service played a particularly important role in providing transport for some Muslim women who, for cultural reasons, were not able to travel alone with male taxi drivers. It also provided a valuable service for a significant proportion of the women interviewed who had a range of needs in relation to disabilities and reported lack of confidence in mainstream taxi services in providing for them. In a similar number of cases women disclosed that they were survivors of a range of personal attacks including rape, harassment and racial abuse. In a larger number of cases women recalled past experiences of travelling with male taxi drivers when they were made to feel threatened.

Although it can be argued that a dedicated service approach detracts from the longer term aim of providing safe transport for all, women-only transport can play an invaluable role in filling the gap in mainstream services. It also provides a means of looking at best practice in safe transport which can be disseminated to commercial taxi companies.

CONCLUSION

Crime has been one of the biggest issues faced by policy-makers in the last decade. Crime-prevention activities have broadened and expanded dramatically over this period and have created exciting debates about the appropriateness and effectiveness of different types of solutions in the prevention of different types of crime. Women's safety has been seen by some commentators as a winner in this burgeoning activity, in the sense that it has been identified as a crime-prevention policy area in its own right and one which needs its own specific interventions.

Women's safety presents a particular challenge in the crime-prevention debate because of the complex nature of women's fear of crime which is grounded in real, but sometimes unquantifiable, experience. The existence of domestic violence also means that solutions which focus on the public sphere, which have a crucial role to play, may nevertheless be seen as not taking full account of women's experience of male violence.

There have been real gains made in the field of domestic violence in the period of growing crime-prevention activity outlined above. Most cities now have established practices of inter-agency co-operation between agencies involved in the support of women who leave violent partners including, amongst others, refuges, housing authorities, health authorities and social service departments. There remain, however, gaps in service provision in terms of longer term support which is not crisis led. There is also very little preventive work taking place.

The case study community-based support scheme outlined in this chapter provides food for thought in considering new and alternative approaches to women's safety. Such locally based schemes, if they are run flexibly, and in accordance with local women's needs, are a potential site for preventative work which can be delivered through education and the provision of practical and emotional support on a short-, medium- or long-term basis. They have a particularly powerful role to play in high-crime and deprived communities where they will often represent the only safe place which local women can inhabit. This role as a 'safe haven' takes on particular importance for women who are experiencing violence at home.

Educating men also has a pivotal role to play in preventing violence against women. There are very few schemes which specifically target violent men and there is little progress in establishing these issues onto a broader educational agenda. Beyond a given level of provision, tensions exist over whether such schemes divert resources from initiatives more specifically targeted at women. It is apparent, however, that in practice what is required is a mixture of schemes: a range of interventions which are different in delivery and in emphasis but which are ultimately complementary to each other. The need for experimentation with different types of intervention is also clearly apparent and, in particular, the need to explore of the role of more social approaches to women's safety.

NOTES

1 The Safer Cities initiative was established as part of the Government's 1988 Action for Cities programme. Its stated aims are 'to reduce crime, reduce the fear of crime and to create safer cities where economic enterprise and community life can flourish' (Home Office, 1984). Safer Cities is intended to promote partnership working in crime prevention and to form an integral part of inner-city regeneration.

2 For full discussion of the findings see Atkins (1989).

9.

WOMEN AND TRANSPORT

Rosalie Hill

INTRODUCTION

Major changes have taken place in the past decade both in the direction of transport policy and in women's social and economic position in society. At just the same time as women are beginning to experience new travel needs and opportunities, it is being recognised that the environment can no longer sustain uncontrolled growth in travel and car ownership. This chapter aims to reappraise the role of gender in explaining travel behaviour and to consider its implications in the new policy environment. The study starts with an exploration of prevailing ideas about women and transport and their origins in past trends. It then draws together recent evidence, from Britain and elsewhere, to assess the extent to which established perspectives remain well founded and considers how far knowledge about women and transport has been both instrumental in bringing about policy changes and in affecting women's transport opportunities. Finally, it draws attention to a new agenda about women and transport which is beginning to emerge in response to evidence of the increasing polarisation of the position of women in society and to global policy commitments towards securing more equitable and sustainable transport strategies.

TRANSPORT ISSUES AND FEMINIST CRITIQUES UP TO 1990

During the 1970s transport planning extended from its established technical base and began to be recognised as an instrument of social policy, whilst administrative reorganisation offered the prospect of integrated transport strategies for the metropolitan areas of England and Wales. Nowhere was this better illustrated than in South Yorkshire, where the radical urban restructuring and capital intensive transport programme of the 1960s, associated with the recommendations of the *Buchanan Report* (Department of Transport, 1962), gave way in the 1970s to a policy of large and indiscriminate bus subsidies aimed at a redistribution of wealth and opportunity (Hill, 1986). With cheap fares

111

available to all over the whole of the South Yorkshire bus network, public transport use was sustained against a declining national trend, but there was little opportunity to target, monitor and respond to the needs of particular social groups, such as women. The prospect of improving access to a cheap, well-integrated public transport system was overturned in the context of a wider post-modern fragmentation of local politics and a retreat from the welfare state in the mid-1980s (Goodwin, 1992). Simultaneously, central government outlawed indiscriminate transport subsidies, abolished the Greater London County Council and the Metropolitan Counties, including South Yorkshire, together with their strategic planning functions, and privatised and deregulated bus operations outside London (Hill, 1993).

Seminal ideas about women and transport in the UK were to emerge in the transition between these two periods, drawing particularly upon the work of Pickup (1984) and the *Changing Places* report of the GLC Women's Committee (GLC, 1985). These studies used contemporary evidence from the 1975–76 and 1978–79 National Travel Surveys and the 1981 Census of Population, together with specially commissioned surveys (TRRL, 1981; GLC, 1985), to characterise women as deprived of access to cars, dependent on public transport, and disabled by their caring and carrying responsibilities.

The underlying assumption of all these studies was that 'women's travel circumstances and their social role are inseparably linked' (Pickup, 1984, p. 62). Thus, women's limited access to cars was held to reflect wider gender inequalities associated with their lower economic status and lack of driving qualifications, as well as with men's monopoly over car use in car owning households. Pickup (1984) observed that less than a third of women compared with more than two-thirds of men held driving licences and concluded that 'while two-thirds of women reside in car ownership with their partner, a number of studies show that these women rarely have a car available for their daily needs, whether they own a licence or not' (p. 61).

Women's socially defined and increasingly multiple roles, involving both domestic and waged labour, were associated with variable, multi-purpose trip patterns, typically combining journeys to school, work and shop, whilst women's primary responsibility for childcare was assumed to impair their mobility and to ally them with other less mobile groups, such as the elderly and the disabled, in terms of their transport requirements (Little, 1994). Moreover, as Morrell discusses in Chapter 8, women's actual or perceived vulnerability to personal attack helped to make the women's movement a vociferous advocate for improvements to personal security in the built — and especially in the transport — environment.

Women's distinctive journey to work patterns were attributed to income and time constraints. Pickup hypothesised that the hard and soft constraints faced by women affected their travel choice and that a woman usually returned to the job market when hard constraints, such as responsibilities for childcare diminished and soft constraints, such as housework and shopping could be rescheduled. Blumen (1994) has brought together evidence from British and other, broadly contemporary, international studies to demonstrate that the number of commuter trips taken by women is similar to that of men but trip lengths are characteristically shorter. She associates short journey lengths with the concentration of working women in inner urban areas; their tendency to more casual, part-time and low-paid employment and their greater involvement in domestic and childcare responsibilities, with women consistently recording significantly less free time than men across a variety of international studies. Short journeys have usually been revealed when measured both by time and by distance, but Blumen reports an interesting exception among minority groups in New York where the commuting time of women is similar to that of men, possibly because of their need to combine school, shopping and journey-to-work trips. Other reasons suggested for the low association between the time and distance travelled by these women included their dependence on slower means of public transport and their tendency to live in more central areas which suffer greater traffic congestion.

An important aspect of feminist thought, building on the work of Castells (e.g. 1978), through that of Huxley (1988), Sandercock and Forsyth (1992), and Bondi (1992), has been that the structure of the built environment and its transport interconnections reflects an historically specific gender division of labour, in which women function as housewives only, rather than responding to the increasing participation of women in the workforce and to changing family and household structures (see Chapter 2). This has created an urban structure ill-adapted to the contemporary needs of women. It has been reinforced by transport planning and provisions which focus on accommodating the peak traffic-flows associated with single-purpose journeys direct from home to work (which are more characteristic of male travel patterns), and by a public transport fares structure which typically penalises aspects which are more characteristic of women's travel patterns, such as the need to break a journey, change a transport mode, return by an alternative route, or to make trips which are irregularly distributed throughout the working week.

The ways women have been socialised into their travel behaviour, and the extent to which women's physical characteristics give rise to distinctive transport needs, have been considered in the literature but have not been widely researched. Hamilton and Jenkins (1989) have pondered whether childhood play with prams as opposed to Dinky Toys might influence adult driving behaviour and further argued that 'while subscribing to the view that the "biological" is socially defined and mediated' nevertheless 'anatomy is destiny' and 'well documented average differences between men and women in height, weight and shape ought to be taken into account in vehicle design' (p. 67). The concentration of cycling among young men, and women's lower propensity to cycle in hilly areas, is well established but it is not clear to what extent this is a product of different male and female physiques; their varying sensitivity to environmental factors, including personal safety and risk-taking; or a reflection of their relative capacities for independent, unencumbered travel.

The importance of women as users of public transport does not appear to have been taken into account in traditional vehicle and station designs (Hamilton and Jenkins, 1989), with their limited space for pushchairs and shopping bags, frequent steps and little attention to personal security. This may reflect the low representation of women in all parts of the transport professions and transport industries, to which recent British correspondence (Starkie, 1995) and research in France has drawn attention. In a total of 150 French agglomerations investigated by Duchene and Pecheur (1995), only ten women were found with political responsibilities for transport and 'women managing the public transport network are exceptions which prove the rule' (p. 127).

It has been further argued that misleading messages affecting women's experience of transport have been promulgated by established techniques of data collection and analysis. For example, the cost benefit analysis model COBA 9, used by the Department of Transport and other Highway Authorities to appraise major highway investments, takes little account of differences in the value of time savings arising from the different financial and time budgets of men and women (Department of Transport, ongoing). Time saved is merely valued in relation to paid-employment and ignores the economic contribution which Pugh (1995), for example, has attributed to women's domestic labour and, as a result, Hamilton and Jenkins (1989) argue 'women bear the cost of particular transport policies and projects while receiving disproportionately few of the benefits' (p. 72). The UK Census of Population records only the mode of transport for the journey to work and the linear distance between home and workplace, while most highway authorities, such as Sheffield, maintain no record of trips other than those crossing the central cordon, and so are likely to under-represent women's trips. Although other studies, including the now ongoing National Travel Survey, which is based on some 10,000 personal travel diaries, record more varied trip patterns, and were important in TRRL's (1981) gender analysis, for example, they are limited in terms of their comparability over time and in their ability to disaggregate information spatially. Furthermore, in many instances the Survey does not report journeys of less than one mile and as a result it is likely to under-represent women's journeys, which are characteristically shorter than those of men. Continuous annual data, in the form of the National Transport Statistics, focus almost entirely on vehicular movements and consider people mainly as accident statistics (Gardiner and Hill, 1996). Apart from the impact which such data may have had on the formation of official policy and perspectives, they may also have given insufficient or inappropriate messages to transport operators about the market position and potential of different groups.

A number of British initiatives on women and transport have been undertaken in response to these critiques (Little, 1994) but most continue to draw on the perspectives and empirical evidence of the early 1980s. The abolition of the GLC withdrew an important resource for research and more recent surveys, such as that on women's response to bus deregulation in West Yorkshire (Hamilton and Jenkins, 1989), have been on a very small scale (30 respondents) or have drawn on the experiences of a self-selected group of women in the transport and related professions (RTPI, 1989). Blumen (1994, p. 239) similarly notes that while much of the international research on which her paper is based has been recently published 'it uses data from the 1970s and 1980s'. She observes that 'the continuation of present urban processes, especially the migration of different types of employment from central cities to the suburbs and edge cities may bear serious implications for the general patterns of metropolitan land use and commuting patterns, and thus will affect the spatial expression of gender differences'. She also acknowledges that 'the social context of employed women at both places, home and work, may have changed since the early 1980s' and 'thus, the consideration of a newer conceptualisation of the daily constraints, social as well as spatial, of both genders should be developed' (ibid, p. 245).

CURRENT GENDER/TRANSPORT RELATiONSHIPS

New evidence is emerging to challenge some of these established critiques on the importance of gender in explaining travel behaviour, especially following the publication of the latest census and national transport statistics in various parts of Europe. Whilst there is little doubt that women, as a whole, continue to have less access to cars, to make more use of public transport, remain less likely to use a car for the work journey and make shorter journey to work trips than men (Department of Transport, 1994), these gender divisions appear increasingly to be mitigated by social, economic and demographic circumstances. For example, analysis of relationships between age and access to cars among

CHANGING PLACES

the older population of Sheffield suggests that, while gender provides a powerful explanation for residence in a non-car owning household, tenure and age are more significant indicators (Gardiner and Hill, 1996).

The latest British National Travel Surveys (Department of Transport, 1994, 1995) reveal that the contribution made by men and women to the total number of trips is similar and has remained almost constant since 1978–9, with women making some 48 per cent of all journeys — that is, on average, about one less trip per week than men. Nevertheless, there remain important gender differences in trip lengths. The 1991–93 National Travel Survey shows that within every age group men, on average, continue to travel significantly longer distances than women, with the most noticeable differences occurring in the age band from 30–59 years (Figure 9.1). Major differences occur in trips for business purposes. The total distance travelled per year for business by all men averaged 2,351 miles in 1991–3, whilst women's average was only 377 miles. Conversely, as shown in Figure 9.2, women account for a higher proportion (some 58 per cent) of all shopping trips undertaken by both the 16–29 and 30–59 age bands and are responsible for a similar proportion (57 per cent) of total miles travelled for shopping purposes,

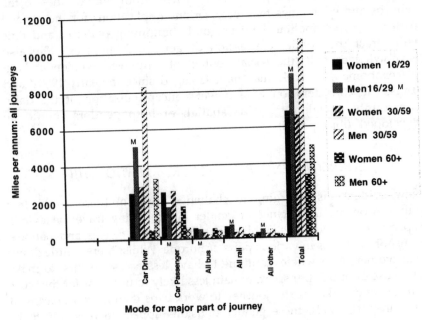

Figure 9.1 *Distance travelled per adult person per year by mode and by age and sex: 1991/93*

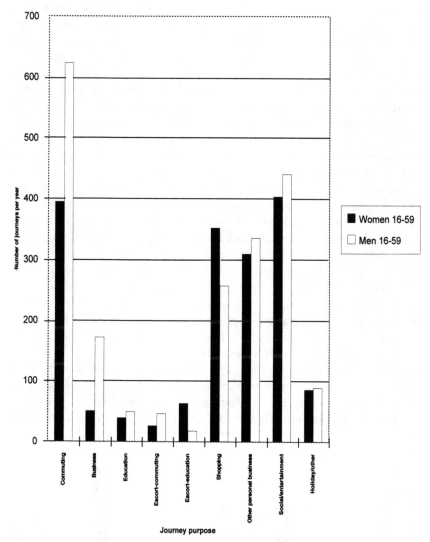

Figure 9.2 *Journeys per person per year by sex for persons 16–59 years*
Source: *Derived from 1989–91 National Travel Survey*

although this represents a major reduction from the position reported by Pickup (1984), drawing on an Oxford study of 1978 which suggested that more than 85 per cent of shopping trips and 71 per cent of escorting trips were attributed to women. In the 30–49 age group women now average 4.4 kilometres per week escorting children (DOT, 1993) — still twice the distance covered by men but Gershuny (1993) has reported interesting variations based on family lifestyles. For example, drawing on evidence from the 1987 ESRC Time Budget survey, he shows that

full-time working mothers with children under the age of two spent only
nine minutes a week in escort duties compared with the 49 minutes
typically devoted to this activity by full-time working mothers of
children aged from 2 to 5 years. Gershuny attributes this surprising
finding to the fact that those women working full-time while their
children are very young are an unusual and relatively high income
group 'which is able to buy in the sort of child care that substitutes for
a typical mother's own escorting time' (p. 68). He further shows that for
working mothers whose youngest child is aged below 5, 'the lower the
income the higher the escorting time, whereas with non-employed
mothers, the higher the income the higher the escorting time'. This he
attributes to escorting having different economic purposes for the
different groups, with low-income earners needing to take a child to a
crèche 'for storage purposes' while high income groups view it as
'helping the child to develop social skills'. Fathers in richer households
are also shown to spend more time in escorting children than those in
poorer households, partly, Gershuny argues, for reasons of 'political
correctness'. Fathers also carry out a much higher proportion of the
relatively smaller number of escorting trips after the working day,
especially those after 9 p.m. — possibly a response to women's fear of
travel at night, and, perhaps, of lesser night-time availability of public
transport for those without access to cars.

Compared with twenty years ago, when the most common method
of transport by women was walking, the most likely mode of transport
for women is now by car, in particular as the car driver (Figure 9.3).
Women car drivers accounted for about a third (34.4 per cent) of all car
trips in 1991–3. Of particular concern is evidence that fear of crime leads

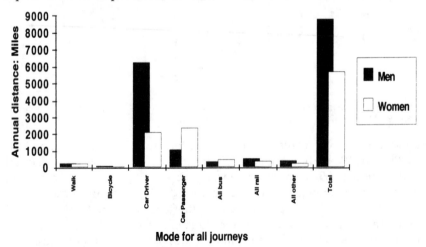

Figure 9.3 *Distance travelled per person per year by mode and sex: 1991/93*

many people, particularly younger women, to avoid public transport and walking (RAC, 1995; Stokes and Taylor, 1995). The latest British Social Attitudes Survey (Pickup *et al.*, 1995) reveals that 3 per cent of men and 8 per cent of women would avoid the use of public transport for this reason, while around a third of people would avoid walking at certain times and 1 in 10 avoided walking whenever possible. More than 1 in 5 of all journeys under a mile are now being made by car, while walking trips have diminished by 16 per cent in the three years up to 1993.

Women drivers are concentrated in the younger age groups, among whom licence holding has been rising rapidly , to the point where it is closely approaching that of men (Figure 9.4). Overall, some 53 per cent of women compared with 81 per cent of men now hold a licence to drive (DOT, 1993), which may be set against the 29 per cent and 69 per cent respectively revealed by the 1975/6 National Travel Survey and widely reported in subsequent studies. Even for women over 60, amongst whom licence holding is still rare, the most common mode of transport

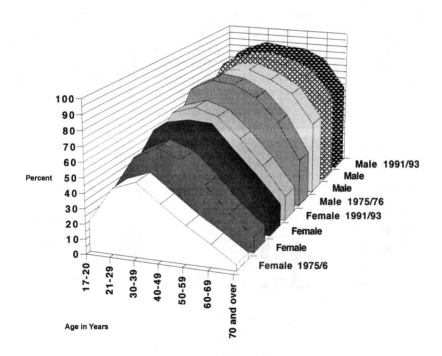

Figure 9.4 *Change in Licence Holding by Age and Sex:* Source: *Derived from 1975/6 and 1991/3 National Travel Surveys*

is as a car passenger. Beuret (1991) argues that the increase in licence holding is still not fully matched by the personal access of women to cars in one car households, especially in the increasing proportion (4.3 per cent in 1991–93) of households which use company cars, although the number of households having more than one car, which has risen from 10 per cent in 1975, 17.2 per cent in 1985–96 to 23.2 per cent in 1991–3 (*Social Trends*, 1995), suggests that many women have increasing opportunities for permanent access to cars.

Nevertheless, despite rising car ownership, some 32.2 per cent of households in Great Britain remain without a car — a relatively modest change from the position two decades ago when 44 per cent of households were without cars (*Social Trends*, 1995). The *Family Expenditure Survey, 1993* (CSO, 1994), shows particularly low rates of car ownership in households comprising one adult and one or more children (which are typically headed by women) compared with those with both a man and a woman in the household — a position little changed since the 1986 survey (Figure 9.5).

The bus is still an important method of travel for many British women — especially for the journey to work — and more than twice as many women as men use bus services (Figure 9.6). Women continue to make fewer rail journeys than men and a rather higher proportion of their journeys by taxi. This may be associated with their characteristically shorter journey distances, especially for business travel, and a concentration of their employment opportunities in central area offices and shops, which are readily accessible by public transport. Personal observation suggests that increasing use of taxis may be being made for shopping trips, which may in turn be a response to more one-stop

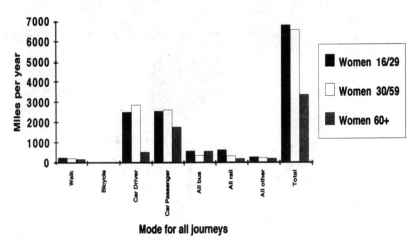

Figure 9.5 *Distance travelled by women per year by mode and age.* Source: *National Travel Survey 1991/93*

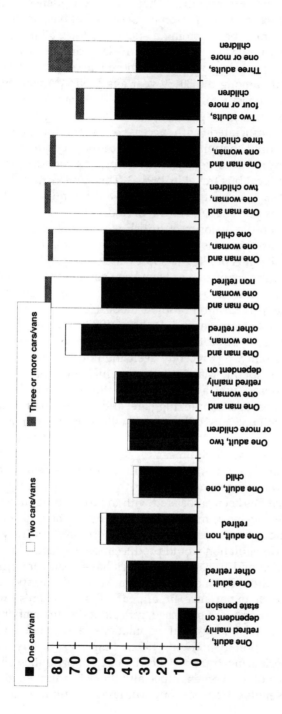

Figure 9.6 *Percentage of households with cars by household composition*
Source: *1993 Family Expenditure Survey*

shopping needs and opportunities. There is some evidence — perhaps in response to concerns about personal safety — that bus, and to a lesser extent, taxi travel may be becoming a substitute for walking by British women, especially in the age groups from 16–29 years and over 60 years. In both these groups, men now average more walking trips over 1 mile than women. Only among the 30–59 year age group do women have a slightly greater propensity to walk than men. This may reflect the likely concentration of caring responsibilities in this age group, and the problems often associated with access to buses for people with pushchairs and bulky shopping. Excluding journey lengths of less than 1 mile, which are not recorded, women accounted for 60.1 per cent of all local and non-local bus travel in 1991–3. This represents only a slight decline from the 63.3 per cent recorded in 1978–9 but must be set against declining total bus patronage over the period.

Thus, rather than the majority of British women being dependent on walking and bus transport, as it appeared in the 1980s, it is now perhaps fairer to suggest that it is public transport which is dependent on women for its patronage. There appears to have been some response to this fact by transport authorities and operators. For example, there is a little evidence that bus privatisation brought modest gains for some women in West Yorkshire (Hamilton and Jenkins, 1989), whilst Sheffield, in addition to its new Supertram system, which was designed for maximum accessibility, has recently seen the addition of some dozen easy-access buses to the Mainline fleet (which dominates the local market). These buses, which can 'kneel' to the kerb and incorporate ramps, wide aisles and ample luggage and wheel-chair space, are observed to be well patronised by women with young children. As well as ramped and lift access, the new Sheffield Interchange has closed circuit surveillance and 'help-lines' to add to a feeling of personal security. However, it is important to set these improvements against an overall diminution in the standard of bus transport, reported by Pickup et al. (1991), which has been associated with the continuing transfer from public to private transport, most particularly by women, in the period since privatisation and deregulation. Furthermore, contrary to the expectation that deregulation would produce more transport innovation, most of the women and transport initiatives have been concentrated in the more regulated environment of London Regional Transport.

From the British experience, it appears that women's transport behaviour is becoming more polarised, reflecting the 'intensifying social and spatial diversity among local populations' associated with post-modern fragmentation (Goodwin, 1992). Some, especially younger, women are experiencing transport needs and opportunities similar to those of men. For other women, especially those heading single-parent and elderly households, transport opportunities remain restricted, while the transport which they can access — such as public transport and walking — has been diminished by the impact of privatisation and growing fears for public safety.

Despite their generally more highly subsidised and better integrated public transport systems than those in Britain, similar trends in women's use of transport are being reported from other parts of Europe. Duchene and Pecheur's (1995) examination of travel in selected French provincial towns of 200,000 to 500,000 population suggests that, as in Britain, men and women do not differ in terms of the total number of trips per day for all purposes, but they do differ in the mode of transport used. In the towns surveyed between 1989 and 1992, men used private cars for 60 per cent of their journeys and alternative modes (walking, two-wheeled vehicles, public transport) for 40 per cent of journeys, compared with a 50/50 split for women. Public transport accounted for 10 per cent of the trips by men and 15 per cent for women, but gender distinctions were more marked in the propensity to walk, which, in marked contrast with Britain, remains high for both genders, accounting for 30 per cent of women's trips compared with 20 per cent of those for men. Further research, of crucial importance to current debates about sustainable urban form, is needed to account for these national differences, which may be related to differences in urban density and land use distribution as well as to different perceptions about personal safety in the built environment.

For employment-related trips, Duchene and Pecheur report that 'the behaviour of women employees is becoming increasingly similar to that of men' with 73 per cent of men and 68 per cent of women using cars, although public transport remains more important for women's journey to work than in their overall travel behaviour. Sixteen per cent of women's against only 8 per cent of men's employment-related journeys were by public transport, suggesting that women may have less access to cars than men for their independent travel; a view consistent with earlier studies, such as GLC (1985), which drew attention to the idea that in single-car households, car use was often dominated by men using it for the journey to work.

Camstra's (1995; 1996) study of commuting in Holland has tested traditional theories which stress women's higher sensitivity to distance and expect them to quit their jobs more often in the case of a residential move. While he found that women did indeed tend to leave their jobs on moving more often than men, this appeared not to be sensitive to the distance moved. Furthermore, women who remained working in the same job after moving house saw their commuting distance increase more substantially than that of men, while 'in the group with the most modern lifestyle' (that is choosing late cohabitation and late child-bearing) 'the gender difference in the percentage of those continuing to work almost disappeared' (p. 283).

Further evidence of polarisation is reported from the United States, where Hanson *et al.* (1995) have found a positive correlation in Baltimore between income and commuting length. This, they argue, has the net effect of creating a vicious circle in which the dual role of women restricts their available journey time and poorly paid jobs locally restrict their chances of increasing mobility.

It is not clear how far the improvement in access to transport for some groups of women (and the increasing polarisation of opportunities) may be attributed to the recognition of gender issues in transport and to the impact of the women's movement. Little's (1994) survey of English local planning authorities conducted in 1991 found that 63 authorities, concentrated disproportionately in the metropolitan areas and London Boroughs, had specific women's initiatives and 33 of these authorities identified initiatives associated with transport. However, there was little variety in the content of the transport initiatives, with the majority specifying the need for greater provision of cheap, safe and accessible public transport. Furthermore, 'at least a third of the initiatives adopted by planners were very broad and generalised . . . dealing with the needs of a wide group of "less mobile" people' (p. 267). Little also notes that 'policies of this kind were not specifically targeted to the particular needs of women, down-playing the issue of safety and frequently failing to recognise the complexity of women's journey requirements' (p. 267). Little concluded from her survey that women's initiatives in general have lost momentum since the 1980s, and that disappointingly few authorities recognised 'the centrality of transport in terms of women's use of the built environment' and the opportunity it provides to 'ameliorate several of the constraints faced by women — such as access to employment and other services, as well as to increase safety and freedom' (p. 267).

A NEW AGENDA

The need for improved understanding of the link between planning and transport — and its impact on women — is central to the Strategy for Sustainable Mobility of the European Union (European Community, 1992). This aims, through a combination of physical, economic and social change, to control the demand for travel and to reduce car dependency in the interests of reducing transport emissions and congestion, while, at the same time ensuring continuing economic development and the integration of all social groups into the mainstream of economic and social life. Through the Eurofem movement, women's issues in transport are being redefined in relation to these wider social and environmental concerns (Ottes, Poventud *et al.*, 1995). The women's movement has always been closely associated with a 'needs-based' approach to transport, aiming for improved access to social, economic and environmental opportunities rather than towards mobility goals, which tend to venerate travel for its own sake and as a symbol of freedom and democracy. Women's restricted time budgets are now viewed more positively as favouring compact urban living and women's less aggressive driving behaviour as being more conducive to fuel conservation. It is being suggested that women's greater experience of public transport and walking makes them the potential best advocates of 'green' modes of transport and that their traditional caring responsibilities create in women a natural concern for the environment and for the welfare of future generations.

However, women, especially those on low incomes and the elderly, constitute one of the few groups among whom there remains scope for major growth in car ownership and there is substantial evidence from recent studies (RAC, 1995; Stokes and Taylor, 1995) as well as in present trends, that women, no less than men, aspire to car ownership and feel dependent on its use. Indeed, it is likely that car manufacturers will positively reinforce women's aspirations by the deliberate targeting of women as one of the few parts of the market with continuing growth potential. Japan, for example, has already seen car companies enticing elderly people with a 'silver car' concept, offering high standards of accessibility and comfort rather than high performance.

These are points overlooked by some commentators who expect a philanthropic response from women, involving acceptance of public transport as a preferred means of transport. A recent experiment carried out by the author among a small group of students suggested just the reverse. When asked to illustrate their ideal mode of transport, the women all focused on individual modes of transport — one a magic carpet, another a covered bicycle, but most looking remarkably like the private car: it was the men who drew trams and trains. To me, this outcome was not unexpected. In many ways, public transport is less easy to adapt to the characteristic transport patterns of women than those of men. It has been shown that women tend to make more complex, encumbered and time-constrained journeys, which are more difficult to cater for economically by public transport than are more concentrated journey to work and business flows. Furthermore, the emphasis on mass public transport, cycling and walking, which is now being promoted is not likely to be well-matched to the transport needs of the growing cohort of very elderly people, predominantly women, among whom mobility impairments are concentrated, nor to those of mothers escorting young children. Indeed, as the difficulty experienced by the new South Yorkshire Supertram in attracting patronage from the low-income areas through which it passes, and a recent study of young people in run-down estates in the Dearne Valley (Casbolt, 1995) both illustrate, many people in the poorer areas of Sheffield, or in the redundant coalfield communities of Yorkshire and Humberside travel relatively little. Since the consumption of travel by any mode rises with income, it is clear that public transport improvements cannot, of themselves, redistribute wealth and opportunity to address the apparent polarisation of women's access to transport. It was recognition of this fact which encouraged Callaghan's Labour administration to challenge South Yorkshire's cheap fares strategy almost a decade before the Conservative government forced its final abandonment (Hill, 1986).

With motoring organisations suggesting that only 20 per cent of journeys could easily be made by modes other than the car and that short journeys under one mile and car escort trips are the most easily discouraged (RAC, 1995), women's travel patterns could too readily become the soft target of policies aimed at reducing car dependency whilst male-dominated business travel, for example, is unaffected. Rather than just looking to improved public transport and safer walking environments as stereotyped responses to women's transport needs, the changing economic power and lifestyles of women might convince manufacturers of the viability of moderately paced, short-range, low-emission vehicles. Such compact electrically, solar or even personally powered vehicles, with more emphasis on flexibility and function than to acceleration and speed, appear quite appropriate to the limited range and diversity of movements which characterise the travel of many women. Modified cycles, with space for shopping and children, rather like those to be found on north European esplanades, for example, could be ideally suited to travel within Sheffield's extensive linear city centre. Community taxis (modelled on the type found in Greece, Turkey and much of Asia) could serve the needs of the elderly and disabled, and provide secure travel for children on the journey to school, while the promotion of car-sharing schemes could also help to bring personal transport into the public domain. Both planners and operators are beginning to realise that bus transport needs to lose its image as a mode of last resort for socially excluded groups and to focus on high quality provision, which meets the widest possible range of access needs and aspirations, including those of women, if it is to compete with the private car.

Further research is required to monitor the impacts of women's changing travel behaviour and their implications for sustainable transport policy. In particular, there is need to examine the relationship between the changing lifestyles and economic empowerment of women and their residential choices and aspirations. What, for example, is the impact on suburbanisation of the different meanings attached to home by men and women, discussed in Chapter 5? Are there also gendered images of transport and what are their implications? What is the impact of women's behaviour on the generation of shopping, escorting and leisure trips, which are now making the greatest contribution to the growth in car use? To what extent will women be prepared to substitute face-to-face contacts with communication on the information superhighway or to relinquish the home/school involvement and school-gate cameraderie which may have been nurtured by a rise in child escort trips? How might those groups of women now approaching retirement with a licence to drive, the experience, of car ownership and greater economic independence, behave in their old age?

The development of an equitable and sustainable transport strategy for the European Union will need to address a similar dilemma to the one faced by Colin Buchanan and the Crowther Commission in Britain in the early 1960s: that is, to what extent is it legitimate and politically acceptable to pull the ladder up on an excluded group who aspire to and have not yet achieved car ownership. Buchanan demonstrated and Crowther advocated a solution which involved adapting the city to the needs of the car and the wholesale reconstruction of environments and communities (Ministry of Transport, 1964). Today, rather than viewing new transport infrastructure as an aid to social and economic regeneration, economic and community regeneration are needed in order to secure sustainable transport strategies based on short range movements. Local economic regeneration can halt the need to extend the job search area. Community regeneration can stem the flight to the suburbs and to better schools, shops and leisure opportunities by making existing settlements more attractive places to live. Only an understanding of the transport behaviour and aspirations of different spatial and social groups — coming from improved opportunities for public participation in transport policy formation — will help us to understand which journeys are most readily substitutable by other modes; and where best to target incentives and resources. The most important legacy of women's concerns about transport from the 1980s will be to emphasise that sustainable transport policy is about meeting people's needs for access and not about providing more and more tracks for vehicles to run on.

10.

WOMEN SHOPPING

Dory Reeves

INTRODUCTION

This chapter sets out to trace the effects trends in shopping have had on the lives of women and how they have been involved in shaping these trends. The chapter starts out by examining some of the terms and then goes on to explore the general literature before looking at the trends and their effects. In strictly economic terms shopping is the process by which people purchase or buy items. A distinction is generally made between essential shopping for day-to-day necessities, the periodic purchase of major household items and what has become known as leisure shopping for the purchase of non-essential items. The terms used to describe different types of shopping acknowledge that it is not just about the 'act of buying' but the whole process leading up to and after the decision to purchase. It involves getting to or accessing shops, having the resources to pay, being able to find something suitable and taking the item to where you live.

For women, shopping can mean a variety of things such as the chore of the family shop, the excitement of a day out, a means of self-expression, exploitative low-paid or flexible work for the majority or a managerial career for a growing number. It will mean very different things to those of different ages and personal circumstances. To a teenager, living at home with few responsibilities for day-to-day needs, shopping means meeting friends, buying disposable fashion, make-up, shoes or jewellery. For those with no money to spend, shopping malls and town centres are places to hang out and people-watch (Anthony, 1985). In employment terms, shops often provide the first paid jobs for teenagers, especially girls, and many will have experienced at first hand the boredom, monotony and low status that can go with the shop assistant's work.

Concepts such as impulse buying, browsing, window, comparative or convenience shopping and bargain hunting are all terms which reflect the trends in shopping. Impulse buying took off with the growth of the self-service supermarket. Browsing is a feature of department-store shopping; window shopping is a feature of the High Street while

128

comparative shopping resulted from the widening choice of shops and products and people's need and desire to get the best value for money. Bargain hunting not only refers to the seasonal sales but to the activities of the car-boot, charity shops and jumble sales sectors. Shop types tend to be associated with particular locations: corner shops with residential areas, department stores with high streets but increasingly with out-of-town stores, superstores and regional centres with major motorway junctions.

The organisation of shopping reflects the way society works, its values, who has power, who takes on what roles and increasingly the use of new technology. The status of shopping as an activity determines the levels of salaries paid to those who work in the retail industry, the opportunities available and who does what shopping. Shop location, design, their relationship to each other, the environment and the facilities provided demonstrate not just the economic decisions being made but how users are perceived.

What is bought reflects not only people's needs and financial position but their chosen lifestyle. As US retail guru Peter Glen put it, 'if you put all your possessions out on the front yard, everything you eat, everything you drive, everything you play and work with — your lifestyle would be there in inanimate objects, and you got them all shopping' (BBC, 1995). The point about shopping is that for many, it no longer bears a close relationship with the satisfaction of basic needs. The growth of consumerism highlights the enormous disparities between those who have a choice and those who don't: single parents, homeless people, those who are long-term ill or unemployed.

The diversity of shops tends not to be reflected in the academic literature which focuses to a large extent on the out-of-town shopping, supermarket and department store sectors, a fact highlighted in a recent review of the geographical and related literature (Jackson and Thrift, 1995). In much of this, the shopping environment is depicted as an impersonal consumer landscape in terms of sites, the links between them and the spaces they occupy with sites then classified into the gentrified areas of the inner city, private comprehensively planned, suburban and ex-urban communities, festival settings and high-tech corridors (Jackson and Thrift, 1995, p. 207).

This limited perspective extends to the shopper. Jackson and Thrift (1995, p. 229) comment that 'there is a paucity of work which extends beyond the Anglo-American [white] frames of meaning' although Jackson and Holbrook (1995) have investigated the different meanings associated with various cultural backgrounds. Bowlby is one of very few academics who have written about women and shopping. Articles in 1984 and 1985 focused on planning for women to shop in postwar Britain. Teenagers and young people hardly figure in the academic literature and when they do it tends to be in the context of social problems such as shop lifting or loitering (Lucia Lo, 1994). Older people have tended to be seen as a disadvantaged group, with Smith's (1991) study revealing that those in the inner urban areas were more dependent on small shops. Although Ware (1992) and McRobbie (1993) have commented on the need to consider gender and racial identities, class and gender identities, the current literature on shopping does not extend to this. The lack of an ethnographic approach to much of the research means that it fails to provide a full picture of all those who engage in shopping. Recent studies commissioned by the Department of the Environment such as that carried out by BDP and OXIRM (1992) on out-of-town retail development and even URBED's (1994) report on town centres have also not considered gender, which is surprising given that women make up the main users of these shopping environments. The Environment Committee Fourth Report on *Shopping Centres and their Future* (House of Commons Environment Committee, 1994) contains only one small section on the consumer. Information needs about retailing tends to focus on local turnover, quantity of floorspace, landuse, transport and environmental data: no attempt is made to disaggregate customers, or to recognise the particular needs of women as regards parking, toilet facilities, rest areas, seating and baby-changing facilities.

The RTPI Report on Planning for Shopping into the 21st Century (RTPI, 1988) was produced by a working party of twelve including one woman representing the Centre on the Environment for the Handicapped. Possibly as a consequence, consumers were defined in relation to the male stereotype; 'working shorter working weeks, will retire earlier . . . will enjoy more leisure time each week and longer holidays, more likely to be a home owner; to live in a two-adult family unit without children with both partners working; and to be a car owner' (RTPI, 1988). Chapter 6 of the report addressed disadvantaged shoppers, defined as low-income earners, residents in areas poorly served by public transport, those without access to a car for routine shopping trips, those with caring responsibilities, the elderly, the disabled, the young and ethnic groups. The more recent Fourth Report of the Environment Committee at least expressed concern for the 'polarisation of shopping behaviour and attitudes between the more affluent households and car owners and the poorer, older and also younger adult shoppers for whom the choice of shopping out-of-town may not be practical' (House of Commons Environment Committee Report, 1994, pp. xiv). Town centre managers are predominantly male while the Association of Town Centre Management represents the retailers, businesses, local authorities and property developers rather than the customers. The British Council of Shopping Centres' publication *Managing the Shopping Environment of the*

Future made scant reference to women as the main consumer group (BCSC, 1987). The consequence of all this is that women have not been adequately consulted as users, experts or customers and their ideas and advice has had few outlets.

The 'new everyday life' approach, dating back to the late 1970s in Sweden, has enjoyed growing influence among women in the rest of Europe (Eurofem Network, 1995). Questioning the functionalist approach to urban planning, it takes into account the way people actually want to live their lives and from this, a much more human, neighbourhood approach is being promoted (Horelli and Vespa, 1994).

The Consumers' Association's (1994) commissioned research undertaken to inform their submission to the above Environment Committee did attempt to present the survey results by gender in recognition of the different types of shopping experienced by women and men. It showed that women and men have broadly the same perceptions of town centres and out-of-town centres with some striking differences. Forty per cent of respondents had no overall preference for where they shopped and they would not want to shop out of town at the expense of town centres. However, Table 10.1 below shows that women are more likely than men to feel safe in a town centre and believe that town centres offer lower prices and a better range of goods. Women are also more likely than men to see the out-of-town centre as a good place for a day out shopping. This complements the finding that they

Table 10.1 How women and men see town and out-of-town centres

Factors	Town centre	Town centre	Out-of-town	Out-of-town	Town Centre/Out-of-town the same	
	Male	Female	Male	Female	Male	Female
	%	%	%	%	%	%
Variety of shops	60	53	20	27	16	15
Choices places to eat and drink	58	52	19	23	14	15
Feelings of personal safety	25	32	19	18	46	40
Cleanliness	20	18	38	35	35	37
Shopping for everyday items	59	67	18	16	16	12
Low prices	28	36	32	24	31	31
Access for people with disabilites	20	23	41	37	19	19
Convenience of opening hours	29	29	33	33	30	28
Ease of getting there by car	21	24	49	40	20	22
Public tranport	64	64	7	8	12	13
Range of goods available	41	45	25	25	26	23
Free parking	14	14	62	58	14	13
Day out shopping	35	36	38	46	12	9
Ease of parking	13	14	61	56	15	13

Source: Taylor Nelson (1994)

are more likely to use the town centre for day-to-day shopping and overall actually prefer town centres more than men.

TRENDS IN SHOPPING

There have been far-reaching trends in shopping which have affected the choices available to women, their employment opportunities and their role in the community. This section provides a brief review of these trends starting with the ubiquitous corner shop, and then goes on to examine some of their effects.

Corner shops, which are still important to many small rural communities, selling every-day items from bread, milk and basic vegetables to tins and toiletries, were a widespread feature of the mid-nineteenth-century shopping environment. Characterised by Hall (1994) as utilitarian and functional with simple fittings, the corner shop for the working class was mirrored by the delicatessen for the middle and upper classes. The twentieth century saw the growth in the small corner shop or general store in the areas of terraced housing and tenements built to house the industrial workers. This was a time when shopping was unhurried and gossipy and has been captured affectionately by writers such as Lilian Beckwith (1971).

The 1960s saw the start of the decline in small shops which accelerated during the 1970s and 1980s with the further expansion of supermarkets stimulated by the growth in car ownership. There has tended to be a focus on the decline of the village shop (RDC, 1994) but the impact on inner urban areas and peripheral estates can be just as devastating (Robertson, 1984). The National Association of Shopkeepers continues to monitor trends and in 1994 reported that six small shops close every day in Britain. These closures are not just a British phenomenon. Clout (1993) recorded a decline in village shops in France where rural car owners were travelling to the new hypermarkets to do their shopping. In the future 'smaller convenience stores in local, suburban catchment areas will reduce in number and will survive only with new capital investment and long opening hours' (RTPI, 1988, p. 30).

The corner shop and newsagent typify many of the conflicts and contradictions shops have for women, combining responsibility for running the shop, running the house and bringing up the family. To women at home, particularly older women and those without access to a car, the corner shop can often be a life-line for basics; the comparatively high price of products being set against the costs of travelling by bus or taxi to the local supermarket.

The biggest development this century in grocery stores must be the introduction of the self-service supermarket. The first of these did not open in Britain until the 1950s although it had been pioneered in 1916 in Memphis through the Piggly Wiggly Stores (BBC, 1995). Key aspects of the shopping experience included the compulsory use of the shopping basket and the maze-like layout of the shelves purposefully designed to slow the customers down. With the self-service store came the growth of impulse shopping. The new stores created a demand for different products and increased customer expectation of choice. Supermarkets have expanded in size and can now be seen in neighbourhoods and town centres, as well as out-of-town centres providing for the bulk shopping of grocery and other items.

Corner shops and supermarkets sell convenience goods — day-to-day items. Department stores in contrast, are very large, selling everything from food and fashion to furniture and cosmetics and are characterised by the separation of goods into different sections, or departments. They still form an important part of the shopping scene today. Designed to create a very comfortable, opulent and luxurious environment, the department stores first appeared in the mid-nineteenth century in France with the opening of Le Bon Marché in Paris in 1869. Macey's in New York followed and Selfridges opened in Britain in 1930. Seen by some as the consumption industry's shop window (Jackson and Thrift, 1995, p. 218) and by others as places for women to socialise and mix, they have been sources of entertainment for all the family, providing services such as childcare, restaurants and reading rooms. They were important in that for the first time the shopper was welcomed into a store without feeling compelled to buy. Stores in the US also provided living accommodation on the premises for young female shop assistants.

Having looked at the history of the Australian Department Store, Reekie (1993) concluded that they created a sexual culture which formulated and reinforced men's power over women. Department stores were designed to exploit what were considered to be women's instinctive and irrational custom by separating them from the more economically rational male by 'sexing' departments. But in a recent review of this book, Judith Allen made the valid point that more could have been said about how female customers responded to the masculine discourses employed in selling since women have often used these stores as meeting places and an escape from the day-to-day chores. Nonetheless, the early US department stores reflected entrenched sexual and racial discrimination and employed only white women on the sales floor. Men received more in-depth training which prepared them for management positions (Dowling, 1993). 'Door-men' were there to help customers in and out of the store but they were also there to create a barrier and help exclude those not deemed suitable to shop there.

As town centres struggled to compete against the out-of-town store, department stores have been seen as dinosaurs by the retail specialist.

But the early 1990s have seen a revival with millions spent on refurbishing stores: Selfridges in Oxford Street, Harvey Nichols in Leeds, House of Fraser in Sheffield, Rackhams in Birmingham and Kendals in Manchester have all been given a new lease of life (*Financial Times*, 1995a). In its annual report, Marks and Spencer (1995) highlights the use of the department store concept in its out-of-town centres to 'show-case' its full range of products. But what facilities have been provided in modern department stores? Lounges are being added for those carrying customer loyalty cards and commercial crèches are becoming more common, but often basic toilet facilities are inadequate and difficult to find.

Whereas department stores provided for the needs of those with middle and upper level incomes, mail order shopping (which also took off in the early part of the century in Britain) provided the means by which working people and hard-up families could buy goods they otherwise could not afford. Referring to a catalogue with pictures and descriptions of items for sale, the shopper makes a selection, then orders direct by mail or through the company's representative. Traditionally, payments have been spread over twenty weeks because the origins of mail order came from the 'shilling' or 'turns' clubs where people would put in a shilling a week and after twenty weeks would in return receive goods worth a pound. Competition between catalogues has always been intense and the 1980s saw a huge increase in specialist mail-order companies. Traditionally women have been mail-order representatives and as such acquired a unique status in the community given their knowledge of people's financial position. They could ensure that a family was able to satisfy its immediate need by paying the first instalment and collecting 'double' next week. In 1995, 49 per cent of the customers of Empire Stores catalogue were in households with an income less than £12,000 per annum (BBC, 1995). With a 10 per cent commission for the representative, running a catalogue provided a supplement to the household income rather than a substitute. Mail order in America, in contrast, enabled those in remote rural communities to benefit from the choice of the city dwellers. The continued growth of mail order is likely and the predictions that by mid-1990s, 10 per cent of customer spending would be by remote ordering appear to have been fulfilled. However, remote ordering is considered by shopping analysts as 'unlikely ever to provide an alternative to a shopping trip for the purchase of most comparison goods, nor can it provide for the "social" and leisure aspects of the shopping trip . . . any growth in mail and other remote retailing methods will bring particular benefits to those consumers who find it hardest to gain access to shopping centres' (RTPI, 1988, p. 31).

Teleshopping, or shopping by phone, using the television or computer as the catalogue, is becoming an established form of remote shopping and will be important for people too busy to get to the shops and parents with small children for whom shopping can be very inconvenient. The distribution networks which have developed to service the mail-order catalogue industry will provide the basis for teleshopping, thus sustaining an important source of employment. However, the jobs of representatives may well disappear, perhaps in the short term, until companies realise the importance of personal contact.

The growth of shopping centres during the 1970s, designed to be covered high streets, integrating shopping with other community facilities such as libraries and health centres, provided a clean, dry and well-lit surroundings albeit with inadequate toilet and crèche facilities. The 1980s saw a time of rampant consumerism for some and destitution for others. At one end of the commercial spectrum, it saw the continued demand by developers for sites for glitzy, state-of-the-art, out-of-town shopping centres and at the other, it saw the growth in car-boot sales, jumble sales and charity shops (Carlsberg and Jenkins, 1992). The large shopping centres do not provide for everyone's needs. For many of the very old, the regional shopping centres are too large, involve a lot of walking and can be very disorientating (Bowlby, 1985). For younger women with driving licences, access to the car will determine where shopping is done. Bowlby (1995) observed that convenience shopping by women often results in little social contact, long queuing and problems in amusing children which make this type of shopping tiring and aggravating. It has only been relatively recently that stores have done much to alleviate this with more check-out staff, children's trolleys and cafés. Studies undertaken as part of the development plan process have highlighted women's concern for the loss of local shops and for the need for adequate toilet facilities in town centres and shopping centres (Sheffield CC, 1993). The late 1980s also saw the growth of low-price or discount grocery shops run by chains such as Aldi and Netto which in effect filled the gap left by the closure of numerous Co-operative stores during the 1970s and 1980s.

Parties, markets, jumble sales, mobile shops and car-boot sales are also important types of shopping for women. Tupperware parties became popular with suburban housewives during the 1960s and still exist, providing a focus for social activities. Street markets in the UK provide a range of low-priced items under one roof, tending to be located just off-centre in cities or in the traditional market square of small towns. In France a quarter of all households visit a market at least once a week (*Financial Times*, 1995b). Mobile shops, the most common being the ice cream van, are under threat. Mobile fish and grocery vans provide a vital service in rural areas and remote residential areas and are particularly important for older people and those without access to cars but are increasingly under threat from superstores which operate special buses to take people to the nearest store.

Traditionally run by women, jumble sales in local halls are generally a means of fund raising but provide important places for people to buy, at very low prices, items such as winter coats and curtains as well as second-hand toys for children. Charity shops often provide a social outlet for retired women looking for employment but who do not need to work for an income, as well as providing work experience for unemployed women and men. Car-boot sales, now a common sight at the weekend, can often be seen cheek by jowl with out-of-town shopping centres (Gregson and Crewe, 1994). Involving products being sold from the backs of cars in an organised venue where a fee is payable for a plot on which to set up stall, they provide opportunities to supplement low incomes from the sale of belongings which are no longer needed and they also provide a hunting ground for those in search of bargains.

SUMMARY AND FUTURE TRENDS

The 1990s has seen the demise of the town centre as a result of recession and competition from out-of-town centres. In 1994, three-quarters of shopping floorspace was still in town centres although the proportion of shopping in out-of-town centres had increased from 5 per cent in 1980 to 17 per cent in 1991 (House of Commons Environment Committee Report, 1994, pp. xix). The decade has also seen a huge growth in catalogue shopping as more people have less time to go shopping. There has been a growth in themed shopping and anti-mall, or small-scale specialist shopping centres, which cater for particular groups of people and a take-off in teleshopping with the growth of cable networks.

Different types of shop provide for the needs of different types of shopper. Figure 10.1 below shows the author's simple conceptualisation of the relationship between women of various ages and the different shop types. It shows that the traditional corner shop provided for the basic needs of women of all ages, whereas other types of shopping are more exclusive and require the person to be over a certain age as with mail order or to be a car user as with out-of-town centres.

In the year 2000 and beyond, shopping analysts speak of people going shopping not for goods themselves but for ideas. Virtual reality stores, it is said, will help us decide how to decorate the home or maintain a healthy lifestyle. Traditional town centres will have to offer more in terms of street theatre, public art and entertainment, refreshments for everyone's needs, to compete with out-of-town centres and this may lead to the continued rebirth of city centres as civic centres.

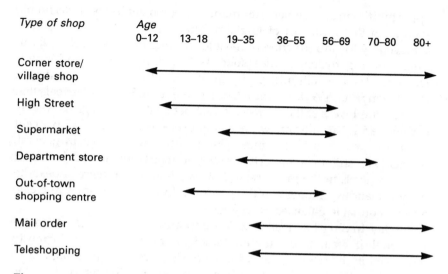

Figure 10.1 *The relationship between shop type and women of different ages*

THE EFFECT OF THESE TRENDS ON WOMEN

The trends in shopping outlined above have affected women on many levels. Economically, shops have provided an important although declining source of employment. Sixty per cent of sales is now controlled by 100 companies, while ten per cent of all jobs are now provided by the retail industry although the retail workforce has declined from 2.9 million in 1961 to 2.3 million in 1993 (Euromonitor, 1995). Women make up about 63 per cent of this figure and the percentage has remained virtually the same for the last 30 years with the majority of jobs being in the large multiples (see Table 10.2 below). Almost half the jobs are part-time compared with just over 28 per cent in 1961 (CSO, 1992); in stores such as Littlewoods and Marks & Spencer the proportion of part-time workers has been found to be over 85 per cent (Penn and Wirth, 1993).

The continued feminisation of shop work has meant increased job

Table 10.2 Structure of the retail industry

	Single outlet	Small multiple	Large multiples	All retailers
Businesses (number)	196,104	22,221	807	219,131
Outlets (number)	196,104	57,806	64,841	318,751
Employees	702,000	282,000	1,340,000	2,324,000

Source: Economist Intelligence Unit (1995)

opportunities in areas where traditional employment has been declining, such as in the north-east of England (Hudson *et al.*, 1992). It has been achieved by the introduction of flexible working practices such as term-time working by companies such as Sainsburys, Asda and Boots. However, there is still a disparity in wages between women and men (MacEwen Scott, 1994). The *New Earnings Survey* (CSO, 1994) reveals that women employees still only receive 72.2 per cent of men's gross weekly earnings and that full-time women employees receive only 79.5 per cent of men's gross hourly earnings. This reflects the fact that in general, women are still in a minority at senior management and board of directors levels in the major retail companies although there appears to be desire and willingness on the part of individual companies to create a more equitable balanced workforce.

Sunday trading has increased job opportunities for some groups of women but created resentment amongst many of those currently employed in the retail trade. A survey of shop workers in 1990, prior to the introduction of Sunday trading, showed that 51 per cent of the sample did not want to work any hours on Sundays and a further 21 per cent only rarely. Some three-quarters of the sample had partners and 72 per cent said that Sunday was the only day when they were able to see them (Kirby, 1993, p. 202).

The gendered nature of shopping and the shopper cannot be denied. The word 'she' is synonymous with grocery shopping (Davies and Bell, 1991) and as Table 10.3 below shows women are still doing the majority of grocery shopping (Pinch and Storey, 1992). By the late 1980s, 85 per cent of women claimed to be in charge of family shopping and today most shopping is still done by women (Woolf, 1994). Women spend more time shopping than men and are more likely to go shopping for non-convenience goods with other women (Sommer *et al.*, 1992, p. 294). As Fiske (1989) said, 'A woman's place is in the mall'. Traditionally, women took men shopping because the man was the income earner. Nowadays, 75 per cent of women have some form of credit card while eight out of ten of them have access to a bank account, whether joint or

Table 10.3 Who does the grocery shopping?

	Men	Women	Shared
All couples	2	59	39
Couple with dependent children	1	61	38
One adult working full-time	2	61	37
One adult working full-time, one part-time	–	61	39
Two adults working full-time	–	59	41
Self-employed	–	74	26
Unemployed couple	9	55	26

Source: Pinch (1992)

single. With the growth of the car and the supermarket, joint shopping was the only way in which most women could get to the shops because fewer women had a driving licence. This is changing gradually but amongst the 50–65 age group, this dependence still exists.

Retailers and policy formers have traditionally been men, whereas employees or shop workers have traditionally been women. The creator of the self-service supermarket, for example, was Clarence Saunders, while the creator of the department store, Le Bon Marché was Aristide Boucicaut. Even the new breed of lifestyle shops developed during the 1960s, which emphasised the importance of design and its accessibility to everyone and not just those with vast sums of money to spend, were developed by men such as Terence Conran who started Habitat in Fulham Road, London. There are a few examples such as Anita Roddick's Body Shop where women have created a new shopping concept and one of the reasons for their small number may be that women have traditionally found it more difficult to get financial support than men.

Women traditionally have not been involved in the formal decision-making process. Instead they have tended to organise around single issues to effect change. Lobby groups such as All Mod Cons have campaigned for the provision of better toilets and public conveniences in shopping and town centres, public transport interchanges and other buildings which the public are encouraged to use from theatres to bingo halls (Greed, 1996). Individuals have led campaigns for decent shopping facilities and after twenty years of lobbying for shopping in peripheral estates like Castlemilk in Glasgow, Josie Livingstone saw the fruits of her labours (Gough, 1994, p. 15). Women planners and architects have campaigned for more facilities in shopping centres and town centres and for effective consultation and the right for women's involvement in the decision-making process. (Matrix Group, 1984) (WDS, 1992) (RTPI, 1989, 1995). A recent paper by Higgins and Davies (1996) documents good practice examples from the UK while Wekerle and Whitzman's book (1995) illustrates examples of how planning, design and management in America and Europe can all contribute to safer cities.

The striking thing about developments in shopping is the limited extent to which the high street reflects the multi-cultural nature of communities. In addition, many towns and cities have not engaged with or tried to respond to the needs of the young. And yet young people have been and continue to be the life-line for city and town centres with the music and clothes shops and the places to meet that satisfy their needs. The out-of-town shopping centres provide a magnet for young people but the private nature of the space in these centres has led to the expulsion of groups of young men and, to a lesser extent, young women. Women with mobility have been attracted to out-of-town centres because they provide under-cover shopping in a pleasant clean environment even though the overall choice of goods may not be any more than a town centre's.

Although excluded from the development process, women have had to accept many of the trends rejecting some innovations (such as the Keedoozle) and adjusting to others (BBC, 1995). The Keedoozle electronic system involved the display of goods behind cases which were released onto a conveyor belt by inserting a key (Kee) which then recorded the costs electronically. One of the big criticisms of this system levied by the early customers of supermarkets was that the system led people to spend more than intended. Women's magazines produced shoppers' guides providing advice on how to counteract this by using a shopping list and keeping track of how much was spent.

Women have traditionally regarded shopping for non-essential items as a day out and as an occasion for meeting friends and relatives, following the fashion trends and escaping from the home work environment. The published statistics however do not show how much time is spent shopping as a leisure activity. A survey of almost 500 female shop workers, most of whom worked part-time, in 37 outlets across Great Britain, found that over half went shopping on their day off as a leisure activity; 47 per cent went on weekdays and 39 per cent went on Saturdays (Kirby, 1993, p. 204). What is clear is that those working full-time have limited opportunities to go shopping for leisure, except at weekends and on holiday. Social Trends (OPCS, 1993), quoting the Henley Centre for Forecasting, showed that women in full-time work had over 10 hours less leisure time a week than similarly employed men, primarily because when they were not at work they spent more time on activities such as looking after the children, cleaning, cooking and essential shopping. This has led some commentators to write about the myth of shopping as a leisure activity as advertised and promoted by the out-of-town shopping centres (Boys, 1990). The development of out-of-town centres accessible principally by car seems to have increased the amount of shopping done by 'couples'. Facilities have consequently been developed for the men who have been 'dragged shopping' by their partners (Cook, 1994). The men's crèche, introduced in the Brand Centre in Enfield, North London, is a sanitised version of a theme pub with Sky television, newspapers, floral wallpaper, wicker arm-chairs, burgundy lamps and a brass ceiling-fan. Located next to the coffee shop it has become a popular resting place not just for men but also for children and women, exhausted by the 'shop-till-you-drop' culture.

Supermarket chains may contend that recent developments in crèches and baby-changing facilities have resulted from their desire to do what they can to make life easier for the shopper, yet for years they appeared to resist pleas from planners and others for crèche facilities and parking spaces (GLC, 1986, RTPI, 1989). Safeway made virtue out of a necessity when in 1992, a crèche was introduced in its new store in Newlands, outside Glasgow, as a result of a requirement by the planning department. Now that stores with these facilities are seen to attract more customers, companies are more willing to comply. On average shoppers who put their child in the crèche spend £10 more per trip (*The Independent on Sunday*, 1994). Over half of Tesco's stores have baby-changing facilities; for Safeway the figure is one in three whilst Asda, with two-thirds of its stores sited in catchment areas with average and below average incomes (ASDA, 1994), has only 13 crèches in a total of 207 stores (Reeves, telephone survey, 1995).

CONCLUSIONS

Every aspect of shopping is still highly gendered from ownership, management, staffing and customers but the review of the literature has shown that there has been a reluctance on the part of academics and policy-makers to explicitly recognise this as a fact although there are some signs that this attitude is changing. The trends in shopping have affected different groups of women in different ways. The trend towards out-of-town centres in the 1980s and early 1990s did not provide for the needs of those without access to cars, including older people and women. The focus on town centres and local shopping centres should help redress this balance but retailers and local authorities need to work closely to provide the facilities needed in shopping areas. The slow-down in the expansion of retail outlets will mean that competition for customers will centre on not just price and efficiency but on the quality of service which can be provided and the facilities made available for shoppers. The growth of teleshopping should help those who are house bound. Many leading stores are gearing up to cater for an exclusive, price-insensitive home market. It is hoped that discount stores will also be able to develop a teleshopping service to provide for those for whom price is all important. Shopping will continue to be important for women both as an activity and as a source of employment and it would appear that the emphasis on part-time work is likely to continue, unless and until the financial advantages of employing part-timers is removed. As consumers, women's influence will continue to be recognised through the growing use by retailers of customer panels. This concept will need to be extended to town centres where male managers have failed to tap into the expertise and advice which could be provided by their largest and diverse group of users.

11.

WOMEN AND LEISURE

Eileen Green

INTRODUCTION

Women writing on the subject of women's leisure have come to the conclusion that it is a contradiction in terms. In-depth studies (Deem, 1986; Green, Hebron and Woodward, 1990; Green, 1996) reveal that for many women the experience of leisure involves both the pleasures of relaxation and the necessity to choose leisure venues carefully, 'guarding' their behaviour within them. It is the quality of the experience and potential companions that are important, rather than the venue or the activity: the chance it provides for a 'break', 'a change', or 'time to be yourself'.

This chapter explores the meaning of leisure for women, suggesting that the traditional work/leisure distinction is inappropriate when applied to women's lives. After summarising recent research on women's leisure, it considers the significance of the relationship between paid and unpaid work for women's access to leisure. The final section explores the issue of women's leisure in the city, raising issues of social control and personal safety which are explored in more detail in chapters 8 and 9. The chapter concludes with a brief consideration of leisure as a potential site of women's empowerment and resistance.

Leisure is defined through ideologies of masculinity and femininity, through ideas about appropriate male and female roles, concerns and behaviour. Traditional academic definitions suggest leisure as 'time free from paid work'; a definition which assumes that everyone has access to periods of 'free time'. Feminist writers (Deem, 1986; Henderson and Bialeschki, 1992) are critical of such definitions, arguing instead that women's heavy involvement in unpaid work and caring responsibilities means that their uncommitted time is at best limited, and at worst non-existent.

WOMEN'S LEISURE ACTIVITIES

Studies of women's leisure in the UK (Wimbush and Talbot, 1988; Green, Hebron and Woodward, 1990), suggest that women's most frequent leisure activities are watching television, reading and home-based crafts, all of which occur in and around the home. These findings are replicated in Australian (Wearing and Wearing, 1988; Dempsey, 1987) and North American studies (Bialeschki and Henderson, 1986; Hunter and Whitson, 1991), suggesting that in western societies at least, women's leisure is more likely than men's to be based in and around the home. Although the 'domestication' of women's leisure is mediated by class and other social and cultural differences, studies continue to show a remarkable convergence between diverse groups of women in relation to the blurring of work and leisure: a convergence which is partly explained by the power of patriarchy.

Biographical approaches to studying women's lives, which rely upon women's diaries as a prime source of data, reveal a surprising uniformity between the lives of celebrated, upper-class feminists such as Virginia Woolf, and Hannah Cullwick, a Victorian working-class woman from rural England. The leisure of both was highly circumscribed by work and men's control of joint and personal income (Stanley, 1988). This remains the case for many women, over a century later (Hargreaves, 1994). Explanations for this are related to women's primary position as mothers and/or carers; free time is a precious commodity which is often taken in snatched spaces between cooking, cleaning and caring for others:

> I class leisure as the time when I've not got (my daughter) because she's my sort of work at the moment, and it's nice to know she's gone to bed and I can do what I want for a couple of hours
>
> (Green, Hebron & Woodward, 1990, p. 6).

Whilst the kind, if not the frequency, of women's recorded leisure activities in the home is similar to those of men, there are clear differences outside the home. This is particularly true of physical recreation in which men participate more than women. Men also engage in a broader range of activities than women. The most popular physical activities for women recorded in the UK (Woodward, Green and Hebron, 1989) were: yoga and keep fit, swimming, badminton, tennis, squash and running or jogging. The tendency to play sport increases with both social class and women's age on finishing full-time education. Middle-class women (a definition based on the male partner's occupation) from households with the highest household incomes in the Sheffield-based study, were twice as likely to undertake some form of physical recreation as those from the lowest income households. Clearly the extent of women's participation in this type of activity is related to their own and others' expectations about appropriate, 'normal' pursuits for women, which in turn are heavily influenced by ideologies of family life and domesticity.

The majority of research on women's leisure has involved white, western women. Much less is known about the effects of ethnicity upon women's leisure. A study of male and female shift workers in the UK (Chambers, 1986), suggests that for this small, qualitative sample of Asian women, home-based leisure was the norm, with 'favourite' activities consisting of dressmaking and other home-based crafts. Turning to out-of-home activities, the Asian women in the sample engaged in only half as many activities as the white women in the study, concentrating on visiting and entertaining friends and relatives and taking holidays and day trips. If we look at the relationship between gender, ethnicity and sport, Hargreaves' work tells us that the bulk of research on black women and sport in the UK has focused upon Asian women, who are statistically the least likely to be involved in sport. Explanations for this include racism, the fact that westernised sports facilities fail to accommodate the cultural and religious needs of Asian women (such as single-sex, enclosed spaces, for example) and the low priority assigned to women's sport by Asian communities. Of all Asian groups, Muslim women appear to be the most restricted in their public leisure; homemaking and domestic roles are highly valued for women and there is an understandable fear of racism and the threat of sexual harassment within Asian communities considering the use of sports venues.

Lesbian women are also subject to discriminatory attitudes and practices which limit their leisure choices. Marginalisation and/or having to pass as 'straight' has led to lesbians establishing separate clubs and facilities as a reaction to the homophobia and alienation they experience when attempting to use mainstream leisure and sports facilities. Women are regularly warned to 'hide' their sexuality in public places, be it lesbian or heterosexual. This fits uneasily with the so-called 'normal' practice of using stylised, fragmented images of women's bodies to sell numerous products, including leisure commodities. The prominent advert for Gossard Wonderbra reminds us that the most desirable women have figures like Barbie dolls or Bay Watch girlies, no matter that these are patriarchal fantasies and not remotely related to the bodies of the vast majority of the female population.

THE WORK/LEISURE RELATIONSHIP

Studies of women's leisure during the 1980s, suggested that access to opportunities and resources for leisure increased for women in employment (Deem, 1986; Green, Hebron and Woodward, 1987a). Being in paid work increases women's financial resources and expands their social networks and in some cases enhances their sense of entitlement to personal leisure (Wimbush and Talbot, 1988). Labour market statistics appear to lend some support to the increasingly popular view that women have taken a major stride towards equality at work. There is clear evidence to support the argument that there is a recent trend towards greater labour market participation among women, and a reduction in men's activity, as discussed in Chapter 2. However, whether this leads to more time for and access to leisure is the subject of some debate.

Firstly, as noted in Chapter 2, women's labour market participation has not increased across the board, the increase being mainly accounted for by married women with partners in employment, and the largest increase is in part-time jobs. Women working part-time are often employed in sectors renowned for low pay and poor working conditions: retail, public services, hotel and catering and cleaning. In addition, women have traditionally been seen as 'flexible' workers, but what has been described as female 'flexibility', has actually always been closely related to women's disadvantaged position in patriarchal society. Furthermore, the flexible worker is being presented as the model for the employment market of the future, where people (women and men) will 'hop' from one job to another (Rosin and Korabik, 1992), becoming a workforce defined by individual 'portfolios' rather than careers in particular types of organisations. While such flexibility and a reduction in hours of paid work potentially offers women the opportunity for increased leisure time, in fact the linked demands made on their time by unpaid caring work, particularly that devoted to adult dependants increasingly marginalised by welfare services, means that most fail to take advantage of this situation by gaining more time for themselves.

There is growing evidence that self-employment for women is increasing. Female self-employment increased by 81 per cent between 1981 and 1989, whereas male self-employment increased only 51 per cent in the same period. Women now represent over a quarter of the total self-employed population in the UK (Carter and Cannon, 1992). Some women are opting out of the traditional labour market for women and setting up their own (small) businesses in an effort to achieve more balance between the demands of employment and family life (Green and Cohen, 1996). On the surface, the reported increase in women entrepreneurs seems to herald an empowerment of women, a taking of control, with a related increase in satisfaction levels and personal autonomy. Once we scratch below the surface, what is actually going on could be seen as a desperate attempt by a minority of (self-selecting) women to gain the flexibility necessary to manage their dual roles more effectively; thereby reducing stress and feelings of guilt and inadequacy.

Although the flexibility of hours gained through becoming self-employed gives women the chance to manage childcare arrangements more effectively and attend key school events, exhaustion was a stark feature of many of the stories we heard in the course of our study, particularly for the mothers in our sample: 'I'm constantly tired, you know. Theoretically I should go back upstairs and do some more work after the kids are in bed, and sometimes I do, but usually I just collapse, brain-dead' (self-employed solicitor, Green and Cohen, 1995, p. 310). Hardly a blueprint for increased leisure time! None the less, the majority of the women interviewed expressed increased satisfaction with their lives since leaving large organisations, which in many cases included more opportunities for time for themselves.

Although it is true that a minority of women are achieving economic security and status through ' breaking the glass ceiling' into the higher echelons of the labour market , they remain a minority (Rees, 1992); the vast majority continue with what has been referred to as 'women's work'. The day-to-day, and lifetime patterns of women's employment indicate that most women take on jobs which 'fit' with other responsibilities as mothers and carers. The increasing proportion of very elderly people in many western societies, coupled with dominant ideologies about women's role being primarily domestic, heralds a situation where the care of elderly dependants could take up more hours of unpaid care than mothering by the turn of the century. Women who have paid jobs, as well as primary responsibility for the smooth running of households, typically find themselves doing what has come to be known as 'the double shift' (Sharpe, 1984; Rees, 1992). The burden of this double shift of housework and emotional caring and employment eats into any available personal space and opportunities for leisure, albeit differently in relation to the age, ethnic identities, social class and sexual orientation of the women concerned.

Despite this, as suggested in the introduction to this chapter, paid work potentially offers increased opportunities for women's leisure. Earning their own money has important advantages for women's sense of self-worth and their authority within the home; however small the wage, economic independence appears to give women a more powerful voice in decision-making in a family context (Hunt, 1980; Pahl, J., 1984). Access to their 'own money', work-mates to go out with and a stronger sense of entitlement to personal leisure means that under the right circumstances , regular employment can actually enrich women's leisure (Wimbush, 1986; Green, Hebron and Woodward, 1987a).

WOMEN'S LEISURE IN AN URBAN ENVIRONMENT

Having explored the meaning of leisure for women, summarised recent research on leisure for women and examined the impact of the complex relationship between employment and unpaid work on women's opportunities for leisure, we now turn a discussion of women's leisure in the city, picking up some of the themes introduced in Chapter 7. As Darke points out there, 'women know that city space does not really belong to them', and yet the city beckons as a place of excitement, pleasure and leisure. Indeed it could be argued that it is the very danger of the city context with its 'sleaze', bright lights and overtones of heightened (hetero)sexuality which draws us there in search of leisure opportunities (Wilson, 1991). Giddens (1990) explores the relationship between time, space and place in modern society, alluding to the social experience of the city, but as Ryan (1994) points out, he fails to engender that experience, thereby assuming a sameness which is fractured by differences of gender, social class and other differences. What is less evident is the role of the urban environment in 'keeping women in their place'.

Returning to the theme of sexuality introduced above, what is clear, is that gender is a crucial dimension in understanding the form and content of leisure in the city in what has been termed 'late modernity' (Ryan, 1994). Ryan argues that the city is a crucial site of social change where the effects of what she refers to as 'the post-modern condition' are embedded in the creation of new urban narratives and social practices. Clearly leisure venues are key sites for the playing out of new social relations in the contemporary city and a study of leisure practices and meanings can reveal rich data on social interaction, not least in the area of gender. Important links are made here between space, place and gender. As Massey argues: 'survey after survey has shown how women's mobility, for instance, is restricted — in a thousand different ways, from physical violence to simply being ogled at or made to feel quite simply "out of place" — not by "capital" but by men' (1994, p. 148). It is at this point that interesting connections can be made between recent work on globalisation, place, time and identity (Adam, 1990; Watson and Gibson (eds.), 1994; Massey, 1994; Urry, 1995) and earlier studies of women's leisure (Deem, 1986; Green, Hebron and Woodward, 1990). Although it is by now well established that time, space and money structured within a capitalist framework, determine our understandings and experience of space, there are other important dimensions which structure our experience, such as race, gender and sexual orientation.

Whether we can walk the streets at night in 'our own communities', venture out of hotels alone in foreign cities, or sample the 'down-town' night life, is heavily influenced by gender and race relations and fear of homophobia. Massey discusses the complexity of what she terms the 'time–space compression' brought about by globalisation. Globalisation is related to economic capital going through a new phase of internationalisation, especially in the area of big business or finance. Clothes are now made in a wider range of countries, increasingly those within what has become known as the 'Pacific rim' and supermarkets are full of food which has been shipped in from all over the world. Our leisure time can now be spent communing with other enthusiasts who share an interest in anything ranging from the work of Michel Foucault to the intricacies of the latest computer software packages; from our sitting-rooms, in front of little screens 'surfing the internet'. There is general acceptance that we now live in a 'global village', which is of course closely related to the rapid development of new information and communication technologies. One of the results of this, according to Massey (1994), is an increasing uncertainty about 'places' and how we relate to them, which is further complicated by what she calls the 'power geometry' of it all. Different social groups and individuals are 'placed' in different but distinct ways, or experience differentiated mobility. There are the so-called jet-setters, 'the ones sending and receiving the faxes and the e-mail, holding the international conference calls. . . . These are the groups who are really in a sense in charge of time–space compression, who can really use it and turn it to advantage, whose power and influence it very definitely increases' (Massey, 1994, p. 149).

But others are more likely to be on the receiving end of it; for example the women whose leisure time is overwhelmingly concentrated within their domestic space, partly because they are 'on call' to the needs of others and partly because of a fear of venturing out alone after dark, especially without safe public transport (Green, Hebron and Woodward, 1987b). Indeed it could be argued that some are imprisoned within time–space compression; mobility is closely related to issues of power and control. Power relations must be understood in relation to gender, as well as in relation to social class and economic capital; especially if we are to understand the links between modernity and changing social relations in the city. The literature of modernity appears to describe the experience of men, since it is about the transformation in the public world; a view of modernity which implies a universal 'disembodied' male, immersed in urban culture.

Urban culture is increasingly globalised due to the effects of mass advertising and the associated growth of popular culture; however it is ironic that the expansion of the 'leisure industry' into what Rojek (1994) (citing Featherstone's (1993) discussion of contemporary male leisure activity) describes as telesex services, should combine disembodied female voices on chat-lines offering sexual narratives 'to stimulate male masturbation', with low-paid, female homeworkers who 're-invent' themselves as blonde 'Barbie dolls'. Whilst alert to the fact that the women (and men) who offer this service to male consumers are low-paid and bear little physical resemblance to the characters they portray, because of the absence of a gender perspective Rojek fails to make the connection between this and the women's (probable) status and experience as female homeworkers, 'imprisoned' within their domestic space, and on call for familial dependants. Feminist analyses of leisure easily make the connection between patriarchal fantasies, and paid and unpaid work for women, unlike some varieties of post-modern discourse.

Gendered analyses of space, place and the city do exist (Massey, 1994; Ryan, 1994) and rightly suggest to us that to understand the positioning of women in the city, we need to understand the historical dimensions of the emergence of urban culture and the gendered place of women within that culture. Walby (1990) outlines critical points in nineteenth-century England when the capitalist demand for labour and the impact of feminist political activity 'fixed' women within a developing, patriarchal city. Ryan (1994) drawing upon the work of Davidoff and Hall (1987), argues that dominant discourses on the propriety of gender-differentiated behaviour in early industrial cities centred on biological differences between men and women. A crucial component of such discourses was the 'sexuality' of women, and their disruptive presence in urban life, which legitimated the construction of separate physical 'spaces' for women and men. Linked to discourses on prostitution, the active sexuality of women was perceived as a disruptive, 'polluting' presence which constituted a threat to public order (Roberts, 1992; Wilson, 1991). The security of ordered family life as represented in the 'private' realm of the 'hearth and home', was threatened by the disorderly chaos of the 'public' world of the outside: on the street, in the city (Davidoff and Hall, 1987).

It was this emerging nineteenth-century domestic ideology which inscribed the place of women within private (female) places and men within public (male) places, and legitimated the ideological exclusion of unescorted women from the city. However, as Ryan (1994) points out, the exclusion of women from the public spheres of the city was not sustained without struggle and did not occur in a uniform way. Ironically, the emerging culture of mass consumption for the middle classes, although it represented new forms of social control over women also created (female) spaces for resistance. The domestic cultivation of 'refinement' in women, which took place in (private) Victorian drawing rooms, also created a space for active female resistance and the building of women's support networks during what were perceived as leisure activities: most of the meetings of suffragettes like the Pankhurst women took place in genteel, middle-class drawing rooms. This domestic ideology of preserving 'normal family life' was to prove a key component of a city culture which excluded middle-class women from public life and regulated the activities of working-class women who by virtue of their occupations regularly frequented the city.

Women were encouraged to exercise proper moral influence in the domestic sphere and became the 'embodiment' of respectability and moral order in public places; a legacy which continues in 'post-modern' societies. Public sanctions for women who challenged the dominant ideologies by appearing unescorted in the city, included loss of social (and sexual) reputation which damaged their marriage chances and public station in society. The loss of social reputation through 'unladylike' activities in a public context was powerfully sanctioned: interestingly enough most often by women themselves, specifically the mothers of unmarried daughters, the details of which are recorded in many Victorian novels. The central text to these discourses was control of women's sexuality; women of all classes and ages were warned to guard their public behaviour against loss of sexual reputation.

Wilson's work makes the link between historical accounts of the urban narrative and contemporary feminist writing on women's leisure in the city. She argues that urban narratives were used to regulate the presence of women in the city; not for them the freedom of the *'flaneur'*, a stroller in the crowd who observes but is not observed. Massey's (1994) feminist critique of the city points out that it is impossible for a woman to be a *'flaneuse'*, because the direction of the *flaneur's* gaze is one-way and often erotic. Since women are always 'bodily objects' of (male) observation, even when they are in tourist mode, the term *flaneur* stands revealed as a gendered concept. For Wilson (1991), the anonymous city provided new opportunities for women to be subject to the sanction of the 'masculine gaze', in that the condition of women became the yardstick for judgments on city life, which ranged from experiences which included forbidden, bodily pleasures, to those associated with moral depravity.

Particularly significant and relevant for our discussion of women's leisure, is the central positioning of the social construction of women's sexuality within narratives of urban life. The development of discourses on sexuality provide the crucial underpinnings of the social regulation of individuals through regulation of their 'embodied' sexuality (Foucault, 1978). Women became and are, a specific focus of this type of surveillance and social control. As argued above, contemporary urban life abounds with stylised, fragmented sexual images of women; as Green, Hebron and Woodward suggest: 'the sexual woman is signified by fragmented bodily parts, most frequently hair, eyes, legs and hips. It is not usually necessary to resort to images of breasts and buttocks when a brief glimpse of a silk-encased knee will convey the message as clearly and more subtly (1990, p. 114).

A key point of interest here is the contradiction between the ways in which women's bodies are used routinely in advertising to represent pleasure, implying that sexuality is a freely available ingredient of the leisure available for consumption, and the fact that the behaviour of 'real' women entering city leisure venues is frequently subject to close surveillance by men; alone, or in groups. The Sheffield study of women's leisure (1987a) (the results of which are corroborated in other international studies) shows us that relaxation and sociability are key components of leisure for the majority of women. For many women, having fun or 'a laugh' in the company of others, outside the home constitutes 'real leisure'; however independent leisure for women can be a source of conflict in couple relationships. This is especially the case when women 'dress up' and enter leisure arenas like pubs and clubs which serve alcohol.

There is now considerable evidence which demonstrates the close link between control of women's sexuality, leisure and male control of public spaces (Hey, 1986; Hunt and Satterlee, 1987; Green, Hebron and Woodward, 1987b). Unescorted women are discouraged from entering what are represented as male leisure venues; they are either made to feel unwelcome or only welcome on specific terms, as, for example, if they are 'waiting for male escorts', or prepared to sit in 'corners', away from the 'male gaze'. Contemporary western society still clings to the premise that (married) women are the property of one man, preferably the husband. By entering certain public spaces without a husband or suitable substitute, women are perceived to have given up their entitlement to his protection and are therefore assumed to be sexually available.

Men's unease about women 'going out' without them is based on a (public) recognition of the above patriarchal norms and codes of practice; this unease intensifies when the chosen venue 'beckons from the glittering city', as the reported behaviour of one of the husbands in the Sheffield study illustrates:

> I've been in a pub and when he's in uniform he's not supposed to come in, you know, come in for a drink — and I've said I'm going in this particular pub in town and he's come in and said: 'Oh, I've just come in for a coke'. Never drunk a coke in his life, you know, but he'll come and have a coke! He's done that twice on me.
>
> (Green, Hebron and Woodward, 1987b, p. 85).

The control of women's sexuality in public leisure venues is linked to issues about personal safety which have been widely researched by feminist writers (Stanko, 1990; Hanmer and Maynard, 1987) and is explored in more detail in Chapter 8. Suffice it to say here that concern for their own safety and that of their women companions constrains women's choice of leisure activities and venues, as does the linked availability of safe transport. The lack of cheap available transport keeps women at home, especially after dark, including middle-class women if they share a car with male partners. As Riseborough and White (1994) argue: 'intimidation as well as actual experience of violence played a part in women's views of the world'; a finding which replicates that of many other studies (Dobash and Dobash, 1992; Edwards, 1987). The Birmingham study found that women's use of space in urban environments was deeply affected by an awareness of danger. Although women's perceptions and experiences of personal safety are affected by diversities of race, class and sexual orientation, there are many commonalities of experience, especially within the context of the city.

CONCLUSION

Many of the themes explored above lead to the conclusion that leisure for women is heavily mediated by lack of money and other resources, friends to go out with and the care responsibilities for dependants. Despite these constraints, and the fact that women's leisure choices are exercised within the limits of patriarchal control, leisure as a site of enjoyment and empowerment is central to women's well-being, as well as being a site of resistance to patriarchal ideologies. Nights out with other women constitute a leisure highlight which is jealously guarded. The importance of women's networks is well documented; freed from the scrutiny of male partners, women are free to 'let their hair down' and engage in 'unladylike' behaviour including ribald humour, which can serve to redress the gendered power imbalances experienced in everyday life.

Girls 'letting their hair down' is very much a feature of the post-modern city as a recent article in the *Guardian* newspaper (Viner, 1994) illustrates. Gangs of girls enjoying themselves at a prime restaurant in London's Hanover Square, tell reporter Katherine Viner that they came without men to have a good laugh: 'God, yes. Men limit you. They're boring. We can say what we like, do what we like — we wouldn't do half this stuff if our boyfriends were here!' Many might argue that women behaving badly are to be avoided, 'they let the side down', but whose side is that? In the words of Katherine Viner: 'The niceness of women has so often been their downfall. It's what keeps them confined to the kitchen, cradle and the patronising arms of men like Kingsley Amis who said, "Women are really much nicer than men. No wonder we like them". No wonder women want to get out and be bad' (Viner, 1994).

Section 4
Changing Places

12.

WOMEN AND CONSULTATION

Chris Booth

INTRODUCTION

This chapter explores consultation with women in the planning process. As well as examining the broader role and development of consultation with women the chapter will also use case studies in Leicester, Sheffield and Birmingham to contrast three different approaches to consultation. The three case studies operate in different institutional, organisational and political contexts; they link with women in the community in different ways and above all the approaches are sufficiently innovative to warrant documentation and bring them into the public domain. Specifically, the case studies seek to focus on process rather than outcomes. They do not provide a blueprint for action, but rather help identify the principles of developing good practice in consultation with women.

THE ROLE OF CONSULTATION WITH WOMEN

The spatial structure, the planning and design of regions, cities, towns and villages, has a profound effect on women's lives. Yet in the past there has been an ignorance and neglect of issues that directly affect women's experience of the environment (Matrix, 1984: Little, Peake and Richardson, 1988). Consultation with women can help raise the profile of issues which are of concern to women and is an important tool in tapping into women's experience of the environment. In addition, consultation of this type can increase our understanding of the interrelationship between issues such as personal safety, childcare, access, mobility, employment, shopping and leisure, which make up the daily pattern of women's lives (see previous chapters).

During the last two decades European society has undergone a number of profound changes. Some of these are demographic, others relate to socio-economic changes. Changing values have made possible a greater plurality of family forms and ways of living. There is therefore a need to research the fine grain of women's, children's and men's lives

(Booth and Gilroy, 1996). Consultation with women can help shed new light on the complexities of daily life and the interrelationships between different activities and different roles.

Little, Peake and Richardson (1988) argue that women's experience of the environment must be seen as a legitimate form of knowledge. The Royal Town Planning Institute Practice Advice Note 12 (1995) *Planning for Women* goes further. It states that 'effective public consultation provides a valuable means of establishing the needs of women in the community', but reminds us that 'needs and priorities will differ amongst women in cities in rural areas, women with children, older women and women from different ethic groups'. Undoubtedly, consultation with women should respect the heterogeneity of women and in so doing value the different experiences and different needs of different groups of women. Using women's experience can help provide detailed knowledge of the daily patterns of women's lives, which in turn can produce more sensitive policies that both recognise the socio-economic change in society and respect the social and cultural diversity of society.

The impact of planning policies on different groups and interests has long been a neglected area, a point noted by the *Development Plans Good Practice Guide* (HMSO, 1992). Consequently, consultation can help assess the impact of traditional planning policy on women. The effect of planning policy on different groups of women has rarely been explored by local authorities and planners. One local authority which attempted to do this was Sheffield City Council. In 1988, the City Council consulted local groups both to provide a client-based perspective on traditional policy formulation in the preparation of its City Centre Plan, and to assess the impact of policy on specific groups using the city centre. More specifically, consultation with women on the City Centre Plan helped evaluate how existing planning policies could disadvantage different groups of women and conversely, how policies could positively benefit women and other groups in the community. In other words, the consultation exercise tried to assess the winners and losers their policies might create.

Consultation also helps women play a role in the management of the environment. The 1990s are beginning to witness the recognition of a plurality of interests in the management of the environment (Healey, 1989, 1990), and it is within this pluralist society that women have a role to play. In many cases, the involvement of women in managing and shaping the environment will be via effective consultation programmes. Jean Augustine at the OECD conference in Paris, 1994 (OECD, 1994), argued that to date 'the management of our towns, villages and cities has been done on behalf of women rather than with women'. Indeed, Booth and Gilroy (1996) have argued for new forms of urban governance based on an inclusive and participatory model, which regard planning processes as a transactional dialogue between different actors and different parties. They suggest that this approach requires the development of more creative participatory techniques, both to enable people to express their needs and to resolve conflicts.

In contrast to many environmental groups contributing to this plurality of interests, women are not an organised group, operating within a prescribed bureaucratic framework. Their organisations are often egalitarian rather than hierarchical, with shifting membership dependent on lifecourse pressures (Foulsham, 1990). Women's interest groups may not even be formalised, rather they can be viewed as a series of overlapping networks that reflect the heterogeneous nature of women in terms of age, class, ethnicity, lifecourse, income, employment and so on. Women's interests can conflict but it is interesting to note that despite these differences women often have similar experiences of the environment and voice similar concerns (Sheffield City Council, 1987; CRESR and SSRC, 1993). Women are not a unified interest group, nor are they highly represented in political parties. However, their lack of cohesion and representation should not be mistaken for a lack of political activity (Bondi and Peake, 1988; Brownill and Halford, 1990). Indeed, they have often been at the forefront of environmental and housing campaigns which directly affect themselves, their families and their homes (Boateng and Wise, 1984; Brownill and Halford, 1990; Hood and Woods, 1994). It is women's experience in the 'informal' political sphere (Brownill and Halford, 1990) which makes them well placed to play an important role in the consultation process.

FRAGMENTATION OF THE PUBLIC INTEREST

In 1968 public consultation was incorporated as a statutory element in the British planning system through the Town and Country Planning Act 1968. At that time the general understanding of public consultation was of an 'homogeneous public' and a 'universal public interest'. Over the last twenty or so years the growth of the women's movement and campaigns for race equality have put the issues of gender and race on the planning agenda. Together, these campaigns have succeeded in fragmenting the public interest (Thornley, 1989), arguably making the nature of the public interest more explicit. Previously 'the public' had been ill-defined and unspecified (Ross, 1991). What subsequently emerged, particularly during the 1980s, was the identification of the community as made up of separate interests, such as women, ethnic minorities and people with disabilities (Booth, 1996). Traditionally these interests have had little say in the way that land and capital has been managed.

During the 1980s there emerged a 'new interest' in planning at the local level, particularly by some local authorities, community groups and other interests (Foulsham, 1990). A few urban labour councils undertook experiments in consultation that aimed to extend power to local people. For example, in the planning arena, local authorities such as the now-defunct Greater London Council (GLC) developed popular planning initiatives; the London Borough of Newham prepared its people's plan for the Borough; Sheffield too prepared its *City Centre Plan for People* (Darke, 1990). All of these local authorities tried to incorporate previously neglected interests, particularly those which were often excluded from the development process: women, people with disabilities, ethnic minorities and the poor. These developments have come to be known as a client-based approach to planning (TCPA, 1986) or popular planning (Darke, 1990).

While many local authorities were undertaking women's initiatives in planning in the 1980s, women's involvement in the planning process was unfortunately not widely documented. That is not to say that consultation did not take place, merely that it went largely unreported. However case studies in Newham, Islington and Lambeth have been evaluated by the TCPA (1986) and Foulsham (1990), and in Sheffield by Darke (1990). These case studies highlight important areas of concern. They reveal a variety of approaches, from formal working parties and advisory groups through to informal discussion groups, questionnaires and day workshops. Consultation in the local authorities studied primarily focused on statutory local plan preparation. The exercises particularly highlight: the need for effective outreach work; targeting of women who are often under-represented in consultation exercises; and involving women who do not belong to formal community organisations (TCPA, 1986). The case studies also reveal the importance of political support and a broad commitment to equal opportunities across the local authority (Darke, 1990; Foulsham, 1990), and they highlight the pivotal role of committed key individuals (TCPA, 1986; Foulsham, 1990). They also support Greed's (1994) view that consultation with women rests on participatory models of bottom-up planning, where the views of women are valued and legitimated in the consultation process. The case studies emphasise that it is the women themselves who best know their areas, their needs and the issues that concern themselves and their families, a view also expressed by the Birmingham for People's Women's Group (1994). Interestingly, effective consultation also revealed the existence of a long-term dialogue between planners and women (TCPA, 1986; Darke, 1990), a feature particularly in evidence in Newham and Sheffield. Most importantly, all the case studies underline the need for consultation with women to respect and to understand the inherent social and cultural diversity of communities and in turn the heterogeneity of women themselves.

WOMEN'S COMMITTEES

Women's involvement in planning cannot be divorced from women's wider access to power and decision-making (Little, 1994). Indeed, these experiments in consultation with women in planning during the 1980s did not take place in a vacuum. Women's initiatives in planning took their lead from the broader movement of feminisation of town halls. The 1970s saw the introduction of the Sex Discrimination Act of 1975 and 1976 and the Equal Pay Act 1970. During the 1970s the attention of the women's movement was firmly focused on central government and on such issues as women's rights and abortion. However, by the end of the 1970s some women in the women's movement began to become active in the Labour Party with a new confidence and determination to raise the profile of women's issues (Lansley et al., 1989, p. 144), and to see women's issues progressed at the political level.

The women's lobby within the Labour party began to challenge the concepts of institutional sexism and the traditional gender-blind approach to policy-making which ignored women's issues and treated everyone alike. In turn, women activists began to re-examine the relationship between women's needs and the delivery of local government services: they challenged employment practices within local government; they put equal opportunities issues on the agenda; and they raised issues of childcare and nursery provision. In 1979, the first women's rights working party was set up in Lewisham. From then on several other London Labour-controlled local authorities followed, most notably the GLC in 1982. Lansley et al. (1989) argue that equal opportunities struck a chord among Labour councils which previously would not have described themselves as being part of the Labour left. By 1987, some 47 women's committees or sub-committees had been established (Lansley et al., 1989). Halford and Duncan (1989), found that by late 1988 some 68 women's 'initiatives' in local government were in existence. More recently Little (1994) undertook a survey of local authorities in England. Although there was only a partial response, and the survey is not directly comparable with the two other studies, Little's survey revealed the existence of only seventeen women's committees and sub-committees, with a further 18 women's initiatives (Little, 1994, p. 83).

Despite the fact that there appears to be no universal pattern or model for these initiatives, it could be argued that during the 1980s there was an emphasis on accessibility, participation and maximum involvement of the maximum number (Lansley et al., 1989). Women's committees and sub-committees broadly attempted to create feminist democracy and incorporate feminist ways of working, and to ensure that women themselves were involved in the decision-making process. For example, women chaired the committees; local women sat on the committees as co-optees; women-only working parties supported the formal committees. The culture and structure of meetings was different; crèches were provided; meetings were held at times convenient to women; and women in the community were surveyed to identify their needs and concerns. Above all, there was a push to draw in different groups of women whose voices were seldom heard, such as black women, lesbians, older women, young women and women with disabilities, and in turn to legitimate and value the experience of these women.

Women's committees often acted as a catalyst to develop women's initiatives in local authority service departments such as planning and housing. Town planning, with its statutory requirement for public consultation provided a useful vehicle to develop more radical consultation strategies and techniques. Women's committees embodied the notion of empowerment and built on the experience and practice in the women's movement.

CASE STUDIES

The three case studies of Leicester, Sheffield and Birmingham provide examples of consultation initiatives operating in contrasting institutional, organisational and political contexts. The initiatives link with women in the community in different ways. The case studies focus on an evaluation of process rather than the outcomes of the initiatives.

Leicester

Leicester's work should be seen as part of a much broader corporate commitment to equal opportunities, and was spearheaded by the City Council's Women's Equality Unit. Since the 1980s the broader work on equal opportunities has enjoyed all party political support in the City Council. Leicester undertook a project to improve women's safety across Leicester (Leicester City Council, 1993a) and the Women's Equality Unit and the community safety co-ordinator successfully applied to the Leicester Safer Cities Project for funding. Work already carried out in Leicester had identified women's safety issues as an important concern. The project had three main stages:

1. the formation of a steering group;
2. consultation with women in Leicester; and
3. presentation of findings and recommendations to appropriate
   service providers.   (Leicester City Council, 1993a)

The consultation process specifically targeted groups of women who were often under-represented in consultation exercises, yet had a high representation in Leicester. The target groups included black women, single parents and women with disabilities (Leicester City Council, 1993a). Women's groups and individual women were identified and contacted through networking with officers and women activists in the community. Nine separate groups were set up to cover a wide range of ages, ethnic backgrounds, locations and lifestages. Six sessions were held with each group. A paid facilitator from outside the local authority was appointed for each group, and meetings were held in locally accessible community facilities, at dates and times convenient to each group. The groups largely set their own agenda and identified issues important to the group. Expenses (such as taxi fares for women with disabilities) were paid to allow women to attend meetings. The facilitator for each group prepared a report of the discussions, and these were subsequently drawn together in a final report. A conference was then organised at which the report and its findings were presented. The conference delegates included some representatives from the local authority, non-local authority organisations, voluntary agencies and some women who had been involved in the consultation process. At the conference a steering group was set up to oversee the implementation of the recommendations in the report. Membership of the steering group included representation from the local authority, other public sector organisations, voluntary groups, women's organisations and private sector companies (Leicester City Council, 1993b). Underpinning the membership of the steering group was a commitment to ensure the representation of a wide cross-section of women's interests. Leicester's work suggests an attempt to change the nature of the relationship between the local authority and the wider community. The consultation initiative has managed not only to involve public, private, voluntary and community organisations, but also to empower those organisations to oversee the implementation of the proposals for improving women's safety in the city.

Sheffield

Sheffield put equal opportunities on its policy agenda in the early 1980s with the support and leadership of its politicians (Ahmed and Booth, 1994). The City Council's Department of Land and Planning pioneered radical approaches to client-based public consultation in the Sheffield City Centre Local Plan in the mid-1980s. The consultation process on the city centre plan involved the use of advisory groups including women, parents with children, people with disabilities, young people, the elderly, the low paid, unemployed people and ethnic minorities (Darke, 1990). Sheffield developed this approach in its work on the Unitary Development Plan (UDP) in 1991 (Reeves, 1995). Building on this work, and coupled with a strong commitment to public consultation, women officers in the Department wanted to set up a mechanism for establishing ongoing dialogue with women in the community. The initial idea to establish a Women's Transport and Development Forum in Sheffield was officer-led and put forward in the Department's 1990/91 Annual Positive Action Report (Sheffield City Council, 1992). During the early 1990s Sheffield was affected by a series of financial crises which succeeded in undermining equal opportunities initiatives through a reduction in staff resources and a reappraisal of work programmes and priorities. During this period support for the forum by senior managers was ambivalent; consequently, the women's forum took nearly three years to set up.

The Women's Transport and Development Forum was finally established in 1993, its main aim being to 'inform the (Land and Planning) Department's work in meeting the needs of women as major users' (Sheffield City Council, 1992). The Land and Planning Department wanted to establish a permanent women's group, which could operate within the local authority framework providing advice on planning and transportation policy, planning applications, and so on. To enable the group to work effectively it was recommended that the group should have no more than 15 members (Sheffield City Council, 1992). 'Membership of the group should represent women of different ages, family life-cycle, ethnicity, class, disability and a cross-section of local opinion' (Sheffield City Council, 1992). Initially, 54 women's organisations were contacted using the data base held for the UDP. By 1995, representatives of 16 groups had attended the bi-monthly meetings, but despite its aims the Forum was largely comprised of white, middle-class, well-educated and middle-aged women, and there were no black women or women with disabilities attending the meetings. The Forum was pioneered by a group of committed officers working in the planning department. The meetings have been held outside of the Town Hall and chaired and serviced by a woman planning officer who has spent approximately one day per month working for the Forum.

Not only have the meetings been used to seek women's views and provide advice for policy formulation, but they also included an element of education and training. In the early days women officers provided women in the Forum with knowledge and information on the role of planning, current planning issues and policies, and the structure and finance of local and central government. Thus training was provided to help increase women's confidence and understanding of the planning process and to enable members to participate in the work of the group more effectively. The group received limited cross-party political support from a few female politicians who lobbied for the group when its future was in doubt. Several female politicians attended the meetings of the Forum, and have taken an interest in specific issues raised in the group, such as planning applications, road safety campaigns and pollution, for example. The Women's Transport and Development Forum fought to be given the legitimacy already offered to other non-elected groups in the Council, such as the Forum for People with Disabilities and the Conservation Advisory Group. The Forum remains officer-led and operates within the local authority structure.

Birmingham

By way of contrast, Birmingham has been a 'bottom-up' community-based initiative outside of local authority control. The Birmingham case study examines the work of a grass roots community-based organisation. The organisation has been organised by women and for women themselves. The Birmingham For People Group (BFPG) is a group of concerned city residents who came together in 1988 to begin to express the needs of all sections of Birmingham residents (Birmingham for People, 1989). The BFPG were concerned at the lack of debate on planning issues in the City (Birmingham For People Women's Group, 1989). Little (1994, p. 192) notes that 'the BFPG is a non-party political group which arose from a national initiative launched by Friends of the Earth and the Greater London Council to advocate the planning of cities round people not cars'. Initially, the organisation developed in response to the planned redevelopment of the 'Bull Ring' in Birmingham's city centre. Subsequently, a women's group was formed to look at how city centre issues affected women. The Birmingham For People Women's Group (BFPWG) comprised a core group of five or six friends who produced a leaflet introducing the issues and asking for further involvement. Another five women came forward to join the group (Little, 1994). This core group wanted to identify women's needs in the city centre and to do this they organised the circulation of 1,000 questionnaires. Two hundred responses were received and a public meeting was also held to seek women's views on the centre. The public meeting was attended by fifty women (Little, 1994). The BFPWG promotional literature states it is a group of women who want a safe, accessible and friendly city for all women, raising awareness of women's planning issues and seeking to encourage a greater input by women into the planning of the built environment of Birmingham (Birmingham For People Women's Group, undated).

As well as producing the booklet entitled *Women in the Centre* (1989), the BFPWG has also produced a video and information pack entitled 'Positive Planning'. The group has undertaken a survey of public toilets in Birmingham (BFPWG, 1993), and has worked with a local community group to produce a safety audit (Safer Estates For Women's Group, 1992). In 1994, the group was working on an information pack on safety and was also involved in setting up an access course in the city for entry into the Architecture and Built Environment professions. The BFPWG sees its role in the community principally as a catalyst for action. The group claims that it is committed to working alongside community groups, gaining women's trust, building confidence, helping collective initiatives and providing advice and organisational skills on women and planning issues (Safe Estates For Women's Group, 1992). The women's group would admit that they have had a membership of young, white, professional single women. To counter this narrow base the BFPWG has networked with other groups, but this has done little to alter the composition of the core group.

The group has received small-scale funding for specific projects from a variety of charitable foundations. Little (1994) noted that the group had formally disbanded due to lack of finance although some of the women are determined to continue to fight for women and planning issues. However, more recent contact with the group suggests that its work continues and that it is very much 'alive and well'. By 1995 the group had been established for over five years with no formal political allegiance or formal political support. Although the Birmingham group has been outside of the local authority decision-making framework, it claims to have sat alongside planners and developers on working parties for major development proposals. The Birmingham For People's Women's Group has campaigned to give women a voice in the future planning of Birmingham. The power and influence of the group is difficult to measure and may only become apparent with hindsight. However, the Birmingham case study suggests that if nothing else has been achieved at the very least women have been given greater visibility in the planning process.

GOOD PRACTICE

This section attempts to identify the principles of good practice that underpin consultation with women. It builds on the author's own experience of consultation with women in South Yorkshire and on the material in the documented case studies. Greed (1994) argues that the women and planning movement has pioneered alternative ways of reaching out to communities. Hood and Woods (1994) examine women's participation in social rented housing through tenant participation, and the barriers women face. It is true that women planners have often been prepared to experiment with more radical approaches: women themselves have been at the forefront of housing campaigns; over the years the feminisation of town halls has provided new models for the democratic involvement of marginalised groups; and women professionals in many of the built environment professions have taken a lead in consultation. However, the principles underpinning much of this work are equally applicable to consultation with men, children, the elderly and other groups in the community. Case studies such as Newham (TCPA, 1986), Sheffield (Darke, 1990) and Leicester (Booth, 1996) provide models of good practice for all practitioners involved in consultation, not just the committed few who are interested in women's issues. Importantly, consultation with women must not be seen as consultation with a special group with special needs. Consultation with women must become a key area of mainstream planning, housing and architecture and it must be seen as part of the wider debate on urban governance.

Effective consultation is more than the use of radical techniques. The consultation process requires a clear framework within which it can take place. The case studies also reveal a number of key principles which can help achieve effective consultation with women.

- *Positive action.* Consulting women does not happen by default; as in other areas of equal opportunities deliberate steps should be taken specifically to involve women in the planning and housing process (Booth, 1996). Consultation with women must be set alongside consultation with other marginalised groups whose voices are seldom heard in the development process. The consultation strategy must identify a cross-section of interests in the community which need to be consulted and then seek to establish a strategy that can successfully reach out to those interests.
- Involving women and other groups in the planning and housing process requires a *redistribution of power* to enable them to influence policy decisions on the management of their environment. The consultation process should seek to empower women and other marginalised groups who historically have had little influence on the decision-making process (Darke, 1990).
- Effective consultation requires widespread *political support* to secure its long-term future (Darke, 1990; Leicester City Council, 1993; Ahmed and Booth, 1994). Support is required for the notion of consultation, the resources required to undertake the process and a commitment to listen to groups in the community and to value their views.
- A *re-evaluation of the traditional role of expert power* in the consultation

process will help develop the empowering professional. The empowering professional can help establish dialogue with women and more importantly can enable women to participate with self-confidence and on equal terms with the expert. The development of the empowering professional will necessitate the experts giving away power, to enable women to take power for themselves.

- The *setting of clear aims and objectives* establishes the scope and purpose of consultation, the interests to be involved and most importantly, defines the negotiable and non-negotiable elements of consultation (Darke, 1990; Reeves, 1995). An awareness of the objectives of the different actors can help build consensus and identify conflicts (DoE, 1994).

- Consultation with women must acknowledge the *social and cultural diversity* of women. Women are not an homogeneous group: we must recognise individual difference as well as those created by social position, cultural heritage, ethnicity, income and disability. Consultation must respect and work with these differences to gain a complete picture of how each of us experiences and perceives the environment.

- *Participatory, bottom-up models of consultation* help involve women on their own terms and in their own environment. Techniques can involve facilitators, focus groups, advisory groups, questionnaires, workshops, outreach work, a forum and so on. Above all, effective consultation employs techniques that do not rely on 'formal' structures and institutional settings, but draw women in to establish dialogue in a non-confrontational manner.

- *Targeting* different groups of women can be useful for tapping into experiences and views that in the past have often been excluded, neglected or simply not heard as, for example, women with disabilities, black women, single parents and so on (Leicester City Council, 1993).

- Consultation will involve elements of *education, training and confidence building*, to enable women effectively to participate in the planning and housing process and to counter criticisms of tokenism (TCPA, 1986; Booth, 1996). All too often women can feel intimidated and lack self confidence (Greed, 1994; Hood and Woods, 1994). The consultation process should seek to empower women, valuing and legitimating their experiences.

- The organisation and implementation of the consultation process must be *sensitive to women's needs*, the various roles they play in juggling the requirements of home, children, work and a family. There must be sensitivity in the organisation of venues, locations, times, dates, childcare provision and access, as well as the use of different techniques to suit different groups of women.

- Effective consultation involves an element of *sustained dialogue and interaction* between professional planners, politicians and women's groups. This enables a relationship to be developed that can help build trust and confidence over a period of time. The empowering professional can facilitate transactional dialogue between different actors and different parties (Ottes *et al.*, 1995). Planners and politicians will also need new skills. What women in the community need is not a process that imposes solutions but facilitators who will listen, debate and respond with sensitivity.

CONCLUSION

As women search for new ways of managing their collective co-existence in shared spaces (Healey, 1994) there is an increasing need to tap into the capacities and skills of all citizens (Booth and Gilroy, 1996). Too many groups have been excluded from the arena of decision-making, and this needs to be corrected not simply to redress the balance but 'to create a different future for all of us, within which neglected and marginalised capacities and perceptions are allowed to flourish and infuse our thinking about what our cities and countryside could be like and how we might collaboratively get there' (Healey, 1994, p. 12). For too long politicians and professionals have been making decisions on behalf of women, now it is time to establish a dialogue with women and to involve women in the management of our towns and cities. Arguably, the new agendas set out in the previous chapters will remain tokenistic without the introduction of new forms of governance.

13.

BREAKING DOWN BARRIERS

Chris Booth

INTRODUCTION

Traditionally, the built-environment professions have been male dominated, yet today these professions are attracting increasing numbers of women. Some professions attract more women than others. For example, in terms of percentages, female representation is greatest in the housing, landscape architecture, town planning and environmental health professions. However, the proportion of women falls significantly in the professions most closely connected with the building and development industry. Table 13.1 and Figure 13.1 illustrate the patterns of women's membership across a range of built-environment professions. It is clear that increasingly women are choosing to enter non-traditional areas and as a result the patterns of membership are changing.

Table 13.1 Women's membership of the built-environment professions*

	% 1995**
Royal Town Planning Institute (RTPI)	23.23
Royal Institute of Chartered Surveyors (RICS)	8.36
Institution of Civil Engineers (ICE)	3.89
Institution of Structural Engineers (ISE)	4.48
Chartered Institute of Housing (CIOH)	51.59
Chartered Institute of Building (CIB)	2.15
Architects and Surveyors Institute (ASI)	2.26
Incorporated Society of Valuers & Auctioneers (ISVA)	unavailable
Royal Institute of British Architects (RIBA)	10.24
Chartered Institution of Building Service Engineers (CIBSE)	1.93
National Association of Estate Agents (NAEA)	21.35
Landscape Institute (LI)	42.38
Chartered Institute of Environmental Health (CIEH)	31.15

Sources: Professional Institute Membership figures as at December 1995

Note:

*In some professions membership of a professional body is a requirement for employment, whilst for others membership is optional. This means that accurate figures for total male and female membership of professions not available.
**The % of membership includes students.

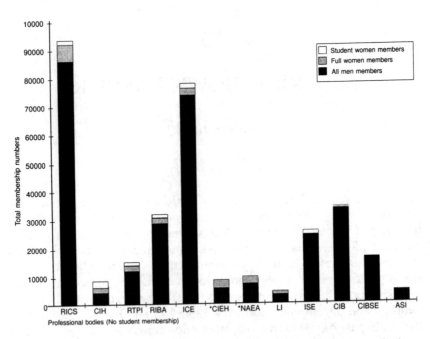

Figure 13.1 *Patterns of Women's Membership of the built-environment professions (1995)*

This chapter discusses the impact of an increasing female presence on these traditionally male-dominated built-environment professions; and more specifically, it questions whether their presence will necessarily change professional values and professional cultures. It also examines the role the professional education process can play both in challenging and reinforcing traditional female stereotypes and indeed how this process can encourage greater female participation in the professions. Organisations and professions can have cultures which set up invisible barriers that serve to maintain the status quo. This chapter provides an overview of the strategies which can be and have been adopted by organisations and individuals for smashing glass ceilings.

Analysis suggests that the built-environment professions can no longer afford to remain dominated by white middle-class males, they need to increase professional diversity to reflect the wider society. The presence of a few token black or female professionals is not enough, there must be a critical mass of women to effect change. 'Malestream' professions require more democratic working environments that value both men and women equally. Ultimately, they will need to create organisational cultures which challenge stereotypes and maximise the potential of all employees.

WOMEN IN THE BUILT-ENVIRONMENT PROFESSIONS

Unfortunately, the increasing numbers of women in the built-environment professions are neither evenly distributed among the various types of jobs, nor are they represented in significant numbers at middle and senior management levels. For example, women working at the professional level in the housing service are concentrated in finance and housing management and not surprisingly are poorly represented in the development field (Brion, 1994). At the managerial level, the representation of women begins to decrease significantly in all three fields, but particularly within development work (Brion, 1994). Nadin and Jones (1990) discovered a similar pattern in their survey of planners. Here too, women planners were least represented in the development industry. In addition, Nadin and Jones found significant differences in salary levels between men and women in comparable age groups. The differences in mean salary levels ranged from around £500 in the 25–29 age group rising to £7,500 in the 50–54 age group. The authors argued that this was partly explained by the fact that more women than men work part-time, but when salary levels are compared between men and women working full-time, similar differentials are revealed. Although this data is limited, not surprisingly, there are indications that job segregation and job stereotyping is very much alive and well in the built-environment professions.

Work on women in male-dominated professions has identified a variety of factors that serve to exclude and marginalise women. Some of the factors include: stereotypes about the nature of women, professions and professionals; lack of female role models, mentors and sponsors; the demand for total commitment from organisations, which may be difficult for women balancing home, family and work; and the exclusion of women from informal activities that help build relationships important for career development (Spence and Podmore, 1987). The authors also point out that the importance of socialisation into gender roles and the domestic division of labour within the home cannot be overstated as factors outside their professional lives which also operate to marginalise women in their working lives (p. 3).

IS MORE BETTER?

An analysis of the membership of professional institutes involved in the built-environment professions, shows that more women than ever are entering those professions, but is more better? Does the presence of greater numbers of women necessarily result in the creation of more gender-sensitive environments and indeed why should it? It is simplistic to assume that should the proportion of women increase in the built-environment professions, then women as a group will necessarily bring a different consciousness to their work. This view assumes that women will do their job differently from men. Although it may be true that some women bring a feminist consciousness to their work, equally others do not, just as some men, but not all, bring politics into their work. This is borne out by Greed's work on women surveyors where she found that for some women gender in the workplace was unimportant (Greed, 1991).

The argument for increasing the numbers of women in the built-environment professions has little to do with women doing their job differently from men or holding different values or indeed changing agendas. An increasingly female presence in the built-environment professions helps challenge traditional female stereotypes and job stereotypes within organisations which are characteristic of male dominated professions (Spence and Podmore, 1987; Little, 1994). In 1983, Audrey Lees, Controller of Planning and Transportation at the Greater London Council (GLC) encapsulated this view when speaking at a North-West Branch Conference of the Royal Town Planning Institute (RTPI). She reminded us, that, 'we are living in a world constructed for men and by men, everything in it being part of this pattern — the hours we work, the holidays we take and even the furniture we use in the office — and yet we attempt to conform' (cited in Bailey, 1983).

As well as challenging stereotypes there is also a compelling argument for the membership of the built-environment professions to reflect the social and cultural diversity of the society they seek to serve. Figure 13.1 illustrates the membership breakdown of the professional bodies by gender. It shows the disparity in total membership and the proportionate lack of women professionals in those bodies. Even in bodies such as the RICS and the ICE, where total membership is between 78,000 and 94,000, the proportion of women professionals remains small. Recently, Professor Michael Romans, President of the Chartered Institute of Building called for greater representation of women, ethnic minorities and people with disabilities in the construction industry. He said, 'To studiously ignore whole sectors of the population by concentrating on white Anglo-Saxon males is foolhardy and hostile to our declared mission' (*The Times*, 6.10.95). The Women's Design Service (undated) note that the real benefits of increasing the numbers of women in fields such as architecture is that this will increase professional diversity and as such help create a profession which more closely represents the society it seeks to serve. Ultimately professional diversity will lead to a greater cross-fertilisation of ideas.

For women and other excluded groups to make their presence felt they need to be present in significant numbers. In other words, there needs to be a critical mass of women. To have one or two professional females within organisations is tokenistic and does little to change the status quo, particularly where appointments are made at a senior level (Ross, 1992). Kennedy (1992) cites Kanter who makes a similar point. She says 'whenever people of any social type are proportionately scarce (i.e. <20% of the total) the dynamics of tokenism are set in motion. Token appointees are more visible and worry about being seen to fail. They are also faced with the choice of accepting comparative isolation or becoming a member of the dominant group at the price of denying their own identity and accepting a definition of themselves as 'exceptional' (Kennedy, 1992, pp. 60–1).

Where women are represented in significant numbers they are more able to provide mutual support and advice; share a female experience of operating in a male-stream environment; and challenge working patterns, practices and local gender cultures. Although increasing the numbers of women in the professions helps challenge professional homogeneity, unfortunately an increase in the number of women, ethnic minorities and people with disabilities is not enough. The values and assumptions upon which the built-environment professions are based can only be truly challenged through the education and training of both women and men. While a more socially and culturally inclusive profession may in the short term do little to change the process or the outcomes, in the long term it can help create opportunities for professionals from a range of backgrounds and life experiences to challenge the traditional values upon which these 'malestream' professions have been built.

CREATING THE MOULD?

We know from previous chapters that women and men lead their lives in different ways. We also know that the lifecourses of women and men are different. Women are diverse and are no more a homogeneous unified group than their male counterparts. Given this diversity, it is not unreasonable to suppose that this reflects itself in their work. Yet this is not always the case as the professional educational process serves to render these differences invisible; irrespective of gender, the process helps create a unified social group of professionals. Observers such as Collins (1990) argue that the education socialisation processes tend to promote similarities within professions, rather than differences. Professionals develop a common language; they may socialise together; they may have a common lifestyle, a code of ethics and a common identity. Because each profession demands that students undergo a common educational process the values promulgated in that process become dominant.

Although women have an increasing presence in the profession, Rydin

(1993) argues they are more likely to take on the values of the men who have dominated the professions for so long. This may be particularly true where there is not a critical mass of women. Greed's (1991) work on women in the surveying profession supports this view. She suggests that it should not be assumed that women entering surveying would necessarily hold different views from men or that they would necessarily be radicalised by being a minority in the profession. This may be partly explained by the backgrounds of entrants and reinforced by the professional socialisation process.

The education process acts as a filtering process controlling entry into the professions. Although Greed (1991) did not specifically research the social class origins of surveyors she found that a considerable number of entrants to the surveying profession came from professional family backgrounds. Similarly, twenty years earlier, Marcus (1971) observed that entrants into the planning profession were predominantly middle class and came from professional family backgrounds. These observations are not surprising as the filtering of entrants through the academic process helps ensure homogeneity within the profession. Yet, today homogeneous professions sit uneasily within the socially and culturally diverse society they seek to serve.

However, observers such as Newman (1993), Coyle (1989) and Cockburn (1991) would argue that women cannot be absorbed into organisations without changing them. There is some evidence that this has already begun to happen. Some women have begun to develop a qualitatively different approach to their work. For example, Matrix, the feminist architectural co-operative, was set up during the 1980s (Matrix, 1984); the Women's Design Service produce advice and offer a resource to women working in planning, housing and architecture; the RTPI has recently published Practice Advice Note 12, *Planning for Women* (Booth and Reeves, 1995); and EuroFEM has created a European network for women working in physical planning and housing (Booth and Gilroy, 1995, 1996).

Many of the professional institutes have already begun to establish women's working parties at national level. At the local level, formal and informal networks have sprung up. For example, the RTPI which has regional women's groups which meet regularly and organise conferences and seminars has delivered a series of training days on skills development in presentation and communication. This initiative in the East Midlands has proved so popular that the meetings regularly attract over fifty delegates. The training days promote an opportunity for women to meet and share experience; they provide positive female role models; and they openly challenge the gender cultures within which women operate.

Feminist analysis has also encouraged some women to question the traditional separation of the personal and the professional. It is interesting that women report that as they go through the lifecourse and ageing process their understanding and awareness of the built environment changes. They have also admitted to using this lived experience and its associated knowledge in their work. For example, in the Polytechnic of Central London's (now the University of Westminster) research project on Women in Planning, a respondent in their survey reported, 'I feel that the knowledge I have gained through bringing up children and doing voluntary work, in addition to my planning background has given me insights into the needs of the local community that I did not have before' (Taylor, 1982). This is a view echoed at women's conferences, in seminars, workshops and tutorials; it is a means of feminising the professions. More women with caring responsibilities work in the built-environment professions and as more men too take on such tasks it will be interesting to observe how far life-course experience will be used by both men and women to enhance knowledge and understanding in the professions, thus blurring the line between professional and personal.

BREAKING THE MOULD?

Nearly all of the built-environment professions have progressively gone through a process of 'credentialisation'. Both practising professionals and those seeking entry to the professions have been encouraged to seek qualifications via an undergraduate or postgraduate route. The direct entry route where people enter the professions as school leavers or with no qualifications appears on the decline. For example, in 1971 as many as 20 per cent of planners directly entered the profession (Marcus, 1971), whereas by 1990 this figure had fallen to 1.9 per cent (Nadin and Jones, 1990). The effect of this process of credentialisation has been to spawn an increase in the number of undergraduate and postgraduate courses, particularly in fields such as planning and housing (Brion, 1994; Greed, 1994). It is important to bear in mind that this process of credentialisation and proliferation of higher education courses is part of a broader social development that goes well beyond the built-environment professions.

Undoubtedly, the professional education process provides a powerful tool for change, but equally it can serve to perpetuate institutional sexism and reinforce stereotypical gender roles. Unwittingly, professional courses can engender discriminatory practices that discourage women's entry and progress within these professions. In Brion's (1994) work on women, education, training and housing, she identified three levels where discrimination against women can operate and it is likely that Brion's findings may be equally applicable to many of the other built-environment professions. The three levels where discrimination can be found are:

1. the *institutional* level where there may be an unequal distribution of men and women in positions of power in education and training. Greed (1994) observed that some planning schools have an average of 10 per cent of women on their staff, although this figure conceals wide variations between Schools. Brion (1994) also notes that in schools of architecture in 1982–83, 97 per cent of teaching staff were male.

2. the *behavioural level* where there is differential treatment of males and females. The Chartered Institute of Housing (CIOH) has had a variety of incidents reported to them which include sexist language, sexual harassment, males dominating discussions, and so on. The CIOH have developed policies and procedures to deal with discrimination, as have some universities. Professions such as the RTPI monitor accredited planning courses with specific reference to equal opportunity issues.

3. the *content level* where the knowledge, values and skills which underpin the course play an important role in establishing the norms in the profession. Professional institutes such as the RTPI have a clear set of educational guidelines which specifically set out the need to take gender issues on board (RTPI, undated). Rydin (1993) also argues that professionals take on the values of the industries and sectors they serve, which in turn are reinforced in the education process.

As increasing numbers of women enter the professional educational process it makes it particularly important that the educational process seeks to remove discrimination and prejudice against women. There are a number of measures that can be taken and professional institutes such as the RTPI have begun to take a more structured and sensitive approach to equal opportunity issues within accredited planning courses. They request data monitoring by gender, ethnicity and disability; they produce educational guidelines on equal opportunities and they disseminate good practice in this field. Summarising literature and research such as WDS (undated), Greed (1991, 1994) and Brion (1994), the work highlights the need for:

• Female role models in the education process and the profession.
• Course content which explicitly takes the issues of gender and the built environment into mainstream teaching to enable all professionals to appreciate that different social and cultural groups have different needs within the environment and so experience the environment differently (see other chapters). The teaching of gender and the built environment must not be seen as the preserve of female

staff; nor should it be marginalised as an option or a specialist subject dealing with 'special needs'.

- Women-friendly courses that positively value the female experience and integrate women students rather than isolate them, particularly where women participants may be in a minority.
- The organisation and delivery of courses to be structured to allow mature women and women with dependants to participate and enter the education process, i.e. appropriate timetables, flexible study routes, location, support structures, crèches.
- Positive action to encourage wider access to professional courses. An increase in the numbers of women entering traditional male-dominated courses will not happen by default. Special efforts must be made to encourage and welcome female applicants, particularly black applicants. Successful campaigns have already been launched by medicine, law and science, so why not the built-environment professions?
- Educational guidelines that clearly set out the required knowledge, values and skills and which acknowledge the significance of gender and the built environment.
- Monitoring of professional courses. This can be done through the collection of data, examination of course material, policies and procedures and discussions with staff and students.
- The adoption of explicit equal opportunities policies within universities. There is a need to eliminate sexism and discrimination in the recruitment of both staff and students; in management and organisation procedures; in course delivery; and in course content and codes of conduct.

SMASHING GLASS CEILINGS

Until now women have been a minority in the built-environment professions. Their situation is similar to that of women in many other professions: they lack status, have lower salary levels, experience job and occupational segregation, and are poorly represented at senior management levels. Traditionally women have had to work in a male world, which has set the norms, values, style and patterns of work. In addition, Maddock and Parkin (1994) argue that women are also subjected to a more personal form of treatment because of their sex. This 'gender culture' of the organisation is difficult to define, but most women working in organisations will recognise the concept. Maddock and Parkin (1996) describe gender culture as: 'men's and women's attitudes towards each other and their interpersonal relations constitute a gendered culture peculiar to each work environment'.

On the basis of work undertaken with British public authorities during the early 1990s, Maddock and Parkin have identified the existence of seven types of gender culture operating in public sector organisations. A summary of their typology is shown in Figure 13.2. As the typology illustrates, gender stereotypes are not the prerogative of men, but can also be perpetuated by women (Maddock and Parkin, 1994). It is likely that gender cultures and gender dynamics are taken for granted within organisations, and being unnoticed set up invisible barriers to the creation of more equal and democratic organisations

Gender culture	Characteristics
Gentlemen's club	Women's role seen as homemaker and mother; man's role seen as breadwinner; polite and welcoming to women who conform; women perform a caring/servicing role; ignores difference and diversity.
Barrack yard	Hierarchical organisation; bullying culture; authoritarian; no access to training and development, clear views on positions within the organisation.
Locker room	Exclusive culture; relationships built on outside sporting and social activities; participation in social sporting activities important to culture of organisation.
Gender blind	Acknowledge no differences between men and women; ignores social and cultural diversity; separation of work from home and life experience; women as perfect mother and super manager; denies existence and reasons for disadvantage.
Smart machos	Economic efficiency at all costs; pre-occupation with targets and budgets; competitive; ruthless treatment of individuals who cannot meet targets.
Paying lip service	Feminist pretenders; Equal Opportunities rhetoric but little reality; espouse views that all women make good managers; women always right; women can empathise.
Women as gatekeepers	Blocks come from women; division between home-oriented women and career women; large numbers of women in support roles; few women in senior positions; women have a sense of place; belief in patriarchal order; pressure on women in senior positions.

Source: Adapted from Maddock and Parkin (1994)

Figure 13.2   Summary of gender cultures

where the capacities of all employees can be maximised. Over the last twenty years, gender cultures have proved powerful barriers to institutional change on equal opportunities (Maddock and Parkin, 1996).

Women and men in the 1990s may increasingly begin to challenge working patterns within the professions as they attempt to achieve a balance between the home, family and work. Abdela (1991) foresees the rise of the 'elder-care' culture and uses information from the Henley Centre, which predicts that by the year 2000 the majority of women in the workforce will be mothers and those caring for elderly and disabled relatives. In her work for the Metropolitan Authorities Recruitment Agency, she argues for what she sees as the 'shevolution'. The 'shevolution' is described as 'the proper balance of men's and women's talents, operating in a work culture adapted to suit both genders comfortably' (Abdela, 1991).

It is clear that many of the built-environment professions and the organisations within which they operate are undergoing a process of rapid change. Therefore, as we move towards the millennium, these professions will need people with appropriate skills; they will need to utilise the potential of all their employees regardless of gender; they will also need to reward both sexes equally. Indeed, Abdela (1991) cites Prime Minister Major, speaking to the National Alliance of Women's Organisations in October, 1990, who said 'we do not as a country have such a profusion of skills and abilities we can afford not to make full use of those we have'.

The term 'glass ceiling' is normally used to describe the invisible barrier that seems to keep women from progressing within organisations. It is important to bear in mind that glass ceilings exist at many levels of the organisation, not just the very top. Barriers can be broken down at two levels; the institutional and the personal. In the past, working at the institutional level has demanded the development of an equal opportunities strategy. Such strategies have sought to challenge and change organisational cultures, norms, values, attitudes, practices and behaviour.

During the 1980s many organisations took up the issue of equal opportunities with varying degrees of success. Abdela's (1991) work identified six key planks in the development of an effective equal opportunities strategy, which are summarised below:

1. A *policy statement* to kick-start the process of change within the organisation. Policy statements must be seen to emanate from, and have the support of senior managers at the highest level. In order to bring people on board it can be helpful to publicise the policy statement both inside and outside the organisation.
2. The development of *systems, structures and guidelines* can help

establish a clear framework for implementation of the strategy. Clearly, without such systems and structures in place, the policy statement will remain empty rhetoric. Such a framework can offer guidelines on issues such as recruitment, promotion, training, language, harassment, working patterns, working environment and so on.

3. *A review of recruitment procedures and promotion targets and practices.* An organisation needs to make itself welcoming and attractive to new female staff, retain existing female staff and develop female staff with the potential for promotion. The review may challenge traditional boundaries between jobs and professions and an organisation may also need to value skills acquired outside the workplace.

4. *Training and development* is an important element in helping people to widen skills and progress in the organisation. Training for female staff can help break down job stereotyping and help women feel confident to move into new areas of work. The use of mentors can also help the career progression of women and may provide a more personal form of development (Morphet, 1992; Coatham and Hale, 1994; McDougall, 1995).

5. *Equality audits* can help the organisation to establish the nature, type and grade of jobs by gender. Audits can reveal the incidence of job stereotyping and the representation of women at management level. Audits are also helpful in examining the culture of the organisation.

6. *Monitoring* provides the organisation with progress checks and takes it beyond lip service to equal opportunities. Presumably an equal opportunities strategy requires resources, and therefore each organisation should be looking for a return on its investment; monitoring helps keep the strategy focused.

Cockburn (1991) predicts that although there is disillusionment with equal opportunities, initiatives will persist and the emphasis will shift from policy-making to policy implementation. She has suggested a number of strategies that have been found to support successful implementation. They include the need to: build a women's movement within the organisation; establish a legitimate difference — i.e. the woman's view; form alliances with supportive men; form alliances with other disadvantaged groups; work with unskilled and part-time women; and build a network of support around the nominated women's officer.

Equal opportunities initiatives have clearly had some success, but they have underestimated the importance of organisational culture as a barrier to change. Progress on equal opportunities has also been undermined by the enormous pressures for organisational change in the public and private sector. More recently, the implementation of equal opportunities in organisations has been described as 'the process of

Managing Diversity' (Kandola and Fullerton, 1994; McDougall, 1995).[1] 'Managing diversity' places gender issues in a much wider context: it is about valuing differences. Managing diversity emphasises the importance of organisational culture rather than the representation of groups within the workforce. Whereas issues of equal opportunities were largely seen as the responsibility of human resource officers, managing diversity places the responsibility firmly with line managers. Equal opportunities focuses on issues of discrimination, particularly on the grounds of race, gender and disability. In contrast managing diversity focuses on maximising the potential of all employees, regardless of gender, ethnicity or disability; it claims to offer a more socially inclusive approach as opposed to an exclusive, separatist approach. Although a welcome development, a word of caution must be sounded on behalf of women who are doubly or triply disadvantaged. It cannot be assumed that black women, lesbians and disabled women will be swept in as diversity is valued.

Equal opportunities strategies have relied heavily on positive action measures, whereas managing diversity quite clearly returns these issues to the mainstream. Some feminists might argue that mainstreaming gender issues will render them invisible and that the issue of women's position in the professions will thus go unrecognised and remain unchallenged. However, equal opportunities strategies have had only limited impact in cracking the glass ceilings, they have also alienated both women and men in some' cases and have not always had the support of the workforce (Coyle, 1989).

McDougall (1995) argues strongly for organisation culture change with an explicit focus on valuing differences between men and women. She acknowledges that an important element in bringing about a change in organisational culture is through the education and training of line managers. She argues for equality to be integrated into mainstream management training. The importance of two further strategies are highlighted for the career development of women: mentoring; and a planned sequence of challenging work assignments or job rotation. Managing diversity is about developing everyone's potential: creating comfortable working environments for both sexes; challenging traditional attitudes and prejudices; and changing organisational and gender cultures that set up barriers for women.

WOMEN AT THE TOP?

It is clear that few women occupy top posts in their professions. Coyle (1989) argues that opportunities for change are limited in local government structures, an area in which many women in the built-environment professions work. Local government is characterised by rigid hierarchies, professional demarcation and a high degree of job segregation. Such structures within local government provide little opportunity for movement from the clerical and administrative grades to the professional and managerial positions. Just as organisations such as local government may set up barriers to opportunity, so do women themselves. Women may suffer anxiety in balancing work, family and caring responsibilities; women can suffer from a lack of confidence in their own experience (Abdela, 1991); and women too hold notions of job stereotyping. In 1988, 11 per cent of managerial staff were women compared to less than 5 per cent in 1971, but they are still most likely to be managers in traditionally female professions (Davidson and Cooper, 1992). So how do women get to the top? Helena Kennedy (1992) observes: 'As in other professions, there is a glass ceiling for women which means that getting to the top floor involves a detour out through the window and up the drain pipe, rather than a direct route along the charted corridors of power' (Kennedy, 1992, p. 58).

At the personal level, the findings by Morphet (1992) and Coatham and Hale (1994) provide us with useful insights into the career paths developed by those women who reach the top. Morphet (1992) identified seven key factors in the successful career development of women, in a study of local authority chief executives. Her findings were substantially supported by Coatham and Hale (1994) in a study of women chief housing officers. A comparative summary of their findings can be found in Figure 13.3.

Coatham and Hale (1994) acknowledge that the studies of the two groups of women at the top in local government displayed some similarities. They also went on to identify a further five factors, which were considered important in the career paths of successful women. These factors go beyond the personal and have wider institutional implications. They include the importance of pre-planning, wide and varied experience, post-entry training and equal access for all, taking risks, childcare and flexible working patterns. It is not surprising that some of the factors identified in the two groups of women are equally applicable to men; they work in 'malestream' professions, where the values and patterns of work have long been established by their male colleagues. Working long hours is a characteristic of male managers and one that can often act as a barrier for women. Indeed few of these women had childcare responsibilities and where they did they minimised their career break; home/job locations may also be similar to that of men. For many women attempting to balance family, home and work, these models of career advancement may be inappropriate. Democratic gender cultures, more flexible working patterns and career paths that value experience and skills gained outside the workplace may

Women chief executives	Women chief housing officers
Pre-planning of career; took a pro-active approach toward career development; 2–3 years in each post.	denial of pre-planning; took a pro-active approach towards career development.
Family support in formative years; father/teacher influential; came from working-class family backgrounds.	Received parental support during early years; came from working-class family backgrounds.
Received the support of a sponsor/mentor from individuals and from the system.	Received the support of mentors inside and outside the local authority organisation.
Had a supportive partner; received encouragement and practical help in combining work, home and family.	Had a supportive partner; received help in running home and family.
Location of home was important; lived close to job; proximity of home useful in working long hours.	No clear pattern; lived on the fringes of metropolitan area where there was a wider choice of job opportunities; home/job location related to partner's career pattern.
Minimised career break; took short break. from work; did not begin career until after children born.	Took short career break or no career break, as majority did not have children.
Hard-working; reported long hours.	Average 10-hour days, worked evenings and weekends; acknowledged workaholics.

Figure 13.3 *Comparative summary of findings on career paths of women chief executives and women chief housing officers. Source: Adapted from Morphet (1992), Coatham and Hale (1994).*

prove a more attractive alternative to the vast majority of women who wish to break through their own glass ceiling.

Studies of the career paths of successful women provide role models as well as strategies for women's career development, but they do not necessarily bring about institutional change. At the personal level, challenge and change rests on the qualities, attitude, strategy and perseverance of the individuals concerned. While breaking down barriers at the personal level is undoubtedly important, and can help challenge stereotypes, it may not always help other women to progress in the profession. It is often tokenistic; women are seen as exceptional, and it can also result in attitudes that suggest, 'she's made it, so why can't other women?'. This is not a helpful scenario when trying to increase opportunities for all women. Most importantly, the development of institutional strategies and changes in organisational culture help to break down barriers. Smashing glass ceilings can create opportunities at all levels which will ultimately result in lasting organisational and societal change.

THE WAY AHEAD?

Women are here to stay in the built-environment professions and their numbers may continue to increase over the next few years. The professions will need to attract and retain people with skills regardless of gender. Increasingly women entrants to the profession will expect to work in all areas of their profession, and at all levels of their organisation. These expectations will inevitably challenge job stereotypes within the professions. Challenge and change wil! necessitate the development of women-friendly professions and organisational cultures that offer comfortable working environments as they attempt to balance home, work and family for both women and men. More importantly, Newman (1993) and Coyle (1989) argue that changes in organisational culture require the involvement of women as agents of change. Coyle (1989, p. 50) observes that 'women cannot simply be absorbed into organisational structures without changing them and it is not so much EO policies which will effect change but the actions of so-called disadvantaged groups . . . who are themselves the agents of future change'.

In practice this means that professions must seek to value both women and men equally; remove barriers that deter women's full participation in the profession; challenge traditional values; widen the representation of entrants to develop socially inclusive professions that are more representative of the society they seek to serve; and finally involve women and men in the process of change. Above all, research needs to be undertaken that goes beyond an audit of women's position in the built-environment professions. Research must seek to examine the experience of women working in these professions in order to provide an agenda for changing places.

NOTE

[1] For a comparison of equal opportunities and managing diversity, see Kandola and Fullerton, 1994.

REFERENCES

AA (1995) *Transport: the Way Ahead — the AA's response*, Automobile Association, London.

Abdela, L. (1991) *Breaking Through the Glass Ceilings*, Metropolitan Recruitment Agency.

Abrams, P. and McCulloch, A. (1976) *Communes, Sociology and Society*, Cambridge University Press, Cambridge.

Adam, B. (1990) *Time and Social Theory*, Polity Press, Cambridge.

Ahmed, Y. and Booth, C. (1994) Race and Planning in Sheffield, in Thomas H., Krishnaryan, V. (ed.) *Race, Equality and Planning Procedures*, Avebury, Aldershot.

Alcock, P. (1993) *Understanding Poverty*, Macmillan, London.

Alexander, S. (1976) Women's work in nineteenth-century London: a study of the years 1820–1850, in Mitchell, J. and Oakley, A. (eds.) *The Rights and Wrongs of Women*, Penguin, Harmondsworth.

Allatt, P., Keil, T., Bryman, A. and Bytheway, B. (1987) *Women and the Life Cycle: Transitions and Turning-points*, Macmillan, Basingstoke.

Allen, S. and Wolkowitz, C. (1986) Homeworking and the control of women's work, in Feminist Review (ed.) *Waged Work: a Reader*, Virago Press, London.

Allen, S. and Wolkowitz, C. (1987) *Homeworking: Myths and Realities*, Macmillan, London.

Anderson, I. and Quilgars, D. (1995) *Foyers for young people: Evaluation of a Pilot Initiative*, University of York Centre for Housing Policy, York.

Anthony, K. (1985) The shopping mall: a teenage hangout, *Adolescence*, Vol. 20, pp. 307–312.

Arber, S. and Ginn, J. (1995) The mirage of gender equality: occupational success in the labour market and within marriage, *British Journal of Sociology*, Vol. 46, no. 1, pp. 21–43.

Asda Group plc (1994) *Report and Accounts for 1994*, Asda, Leeds.

Atkins, S. (1989) *Critical Paths: Designing for Secure Travel*, Design Council, London.

Austerberry, H. and Watson, S. (1983). *Women on the Margins: A Study of Single Women's Housing Problems*. Housing Research Group, The City University, London.

Bailey, C. (1983) Learning to Get by in a Man's World, *Planning 543*, 4 November.

Barrett, M. (1980) *Women's Oppression Today: problems in Marxist Feminist analysis*, Verso, London.

183

Bart, P.B. and Moran, E.G. (eds.) (1993) *Violence Against Women: The Bloody Footprints*, Sage, London.

BBC (1995) *Shopping, The Story of a National Pastime from Both Sides of the Counter*, Transcriptions, BBC Television, London.

BCSC See British Council of Shopping Centres.

BDP and Oxford Institute of Retail Management (1992) *The Effects of Major out-of-Town Retail Development*, HMSO, London.

de Beauvoir, S. (1972) *The Second Sex*, Penguin, Harmondsworth.

Beckwith, L. (1971) *About My Father's Business*, Hutchinson, London.

Beechey, V. (1977) Some notes on female wage labour in capitalist production, *Capital and Class*, Vol. 3, pp. 45–66.

Beechey, V. (1978) Women and production: a critical analysis of some sociological theories of women's work, in Kuhn, A. and Wolpe, A. (eds.) *Feminism and Materialism*, Routledge & Kegan Paul, London.

Beechey, V. (1987) *Unequal Work*, Verso, London.

Benston, M. (1969) *The Political Economy of Women's Liberation*, Monthly Review, Vol. 21 no. 4, pp. 13–27.

Bernard, Y. (1991) Evolution of lifestyle and dwelling practices in France, *Journal of Architectural and Planning Research*, Vol. 8, pp. 192–202.

Beuret, K. (1991) Women and transport, Chapter 4, pp. 61–75, in Maclean, M. and Groves, D. (eds.) *Women's Issues in Social Policy*, Routledge, London.

Beveridge Report (1942) *Social Insurance and Allied Services*, Cmnd 6404, HMSO, London.

Bialeschki, M.D. and Henderson, K. (1986) 'Leisure in the Common World of Women', *Leisure Studies*, Vol. 5, pp. 299–308.

Binney, V., Harkell, G. and Nixon, J. (1981) *Leaving Violent Men: A Study of Refuges and Housing for Battered Women*, Women's Aid Federation England, London.

Birmingham For People (1989) *What Kind of Birmingham? — Issues in the Redevelopment of the City Centre*, Birmingham For People, Birmingham.

Birmingham For People's Women's Group (1990) *Women in the Centre, Women, Planning and Birmingham City Centre*, Birmingham For People's Women's Group, Birmingham.

Birmingham For People's Women's Group (1993) *Caught Short in Brum — Toilets for Women in Birmingham City Centre*, Birmingham For People's Women's Group, Birmingham.

Birmingham For People's Women's Group (1994) *Positive Planning Video*, Birmingham For People's Women's Group, Birmingham.

Birmingham For People (undated) *Towards a Better Bull Ring — A People's Plan*, Birmingham For People, Birmingham.

Birmingham For People's Women's Group (undated) *What is the Birmingham For People Women's Group?*, Birmingham For People's Women's Group, Birmingham.

Birmingham For People's Women's Group (undated) *Women's Access to Architecture and the Built Environment*, Birmingham For People's Women's Group, Birmingham.

Blumen, O. (1994) Gender differences in the journey to work, *Urban Geography*, Vol. 15, pp. 223–245.

Boateng, P. and Wise, V. (1984) *Planning for Equal Opportunities in a Large Metropolitan Area*, Greater London Council.

Bondi, (1992) Gender Symbols and Urban Landscape. *Progress in Human Geography*, Vol. 16 no. 2, pp. 157–71.

Bondi, L. and Peake, L. (1988), Gender and the City : Urban Politics Revisited, in Little J., Peake L., Richardson P. (eds.) *Women in Cities*, Macmillan, Basingstoke.

Booth, C. (1903) *The Life and Labour of the People of London*, Macmillan, London.

Booth, C. (1996) Gender and public consultation: case studies of Leicester, Sheffield and Birmingham, *Planning Practice and Research*, Vol. 11, no. 1, pp. 11–20.

Booth, C. and Gilroy, R. (1995) Changing agendas, *Planning Week*, Vol. 3, no. 35, 31 August.

Booth, C. and Gilroy, R. (1996) Dreaming the possibilities of change, *Built Environment*, Special Issue, Vol. 22, no. 1, Spring, pp. 72–82.

Booth, C and Reeves, R. (1995) Creating opportunities for choice, *Town and Country Planning*, August.

Boston, S. (1980) *Women Workers and the Trade Unions*, Davis-Poynter, London.

Bott, E. (1957) *Family and Social Network*, Tavistock, London.

Bowlby, S. (1984) Planning for women to shop in post-war Britain, *Environment and Planning D: Society and Space*, 2, pp. 179–190.

Bowlby, S. (1985) Shoppers' needs, *Town and Country Planning*, Vol. 54, pp. 219–222.

Boys, J. (1984a) Women and public space, in Matrix Group, op. cit.

Boys, J. (1984b) Is there a feminist analysis of architecture? *Built Environment*, Vol. 10, no. 1, pp. 25–34.

Boys, J. (1990) Women and the designed environment: dealing with difference, *Built Environment*, Vol. 16, no. 4, pp. 249–256.

Bradley, H. (1989) *Men's Work, Women's Work: a Sociological History of the Sexual Division of Labour in Employment*, University of Minnesota Press, Minneapolis.

Bradshaw, J. and Holmes, H. (1989) *Living on the Edge: a study of the living standards of Families on Benefit in Tyne and Wear*, Tyneside CPAG, York.

Brailey, M. (1986) *Women's Access to Council Housing*, Glasgow: Planning Exchange Paper 25.

Braybon, G. and Summerfield, P. (1987) *Out of the Cage*, Pandora, London.

Brion, M. (1994) Snakes and ladders? — women and equal opportunities in education and training for housing, in Gilroy R and Woods, R. (eds.) *Housing Women*, Routledge, London.

British Council of Shopping Centres (1987) *Managing the Shopping Environment of the Future*, BCSC.

Broome, J. and Richardson, B. (2nd edn, 1995) *The Self-Build Book*, Green Earth Books, Dartington.

Brown, J. (1989) *Why Don't they Go To Work? Mothers on Benefit*, SSAC Research Paper No. 2, HMSO, London.

Brownill, S. and Halford, S. (1990) Understanding women's involvement in local politics: how useful is a formal/informal dichotomy? *Political Geography Quarterly*, Vol. 9, no. 4, October, pp. 396–414.

Bruegel, I. (1979) Women as a reserve army of labour: a note on recent British experience, *Feminist Review*, Vol.3, pp. 12–23.

Burghes, L. (1980) *Living from Hand to Mouth*, FSU/CPAG, London.

Burnett, J. (1986) *A Social History of Housing 1815–1985*, Methuen, London.

Callender, C. (1992) 'Redundancy, unemployment and poverty', in Glendinning, C. and Millar, J. (eds.) op. cit.

Cameron, A. and Kuhrt, A. (eds.) (1983) *Images of Women in Antiquity*, Croom Helm, London.

Campbell, B. (1993) *Goliath*, Methuen, London.

Camstra, R. (1995) Gender, household relocation and commuting distance to the workplace: a life style approach, Ch. 2, pp. 11–21 in Ottes, L., Poventud, E. *et al.* (eds.), (1995) *Gender and the Built Environment: Emancipation in Planning, Housing and Mobility in Europe*, Assen, Van Gorcum, The Netherlands.

Camstra, R. (1996) Commuting and Gender in a Lifestyle Perspective, *Urban Studies*, Vol. 33, no. 2, pp. 283–300.

Carlsberg, N. and Jenkins, P. (1992) *Thrift Shopping in England*, Carlsberg Press, London.

Carter, S. and Cannon, T. (1992) *Women as Entrepreneurs*, Academic Press, London.

Casbolt, D. (1995), Listen to Young People, The Children's Society.

Castells, M. (1978) *City, Class and Power*, Macmillan, London.

Central Housing Advisory Committee (CHAC) (1969) *Council Housing: Purposes, Procedures and Priorities*, HMSO, London.

Central Statistical Office (1992) *Employment Gazette*, HMSO, London.

Central Statistical Office (1994) *New Earnings Survey*, HMSO, London.

Central Statistical Office (CSO) (1995a) *Social Trends*, HMSO, London.

Central Statistical Office (CSO) (1995b) *New Earnings Survey*, HMSO, London.

Central Statistical Office (1996 edition) *Social Trends*, HMSO, London.

Chambers, D. (1986) The constraints of work and domestic schedules on women's leisure, *Leisure Studies*, Vol. 5, pp. 309–325.

Chua, B-H. (1991) Modernism and the vernacular: transformation of public space and social life in Singapore, *Journal of Architectural and Planning Research*, Vol. 8, pp. 203–221.

Clark, A. (1982) *Working Life of Women in the Seventeenth Century*, Routledge & Kegan Paul, London.

Clatterbaugh (1990) *Contemporary Perspectives on Masculinity*, Westview Press, Colorado.

Clout, H. (1993) Retailing in Rural France, in *Geographies et Campagnes*, pp. 161–169.

Cm. 1264 (1990) *Children Come First: the Government's Proposals on the Maintenance of Children*, HMSO, London.

Cm. 2901 (1995) *Our Future Homes. opportunity, choice, responsibility*, HMSO, London.

Coatham, V. and Hale, J. (1994) Women achievers in housing: the career paths of women Chief Officers, in Gilroy, R. and Woods, R., op. cit.

Cockburn, C. (1983) *Brothers: Male Domination and Technological Change*, Pluto Press, London.

Cockburn, C. (1991) *In the Way of Women, Men's Resistance to Sex Equality in Organisations*, Macmillan, Basingstoke.

Cohen, L. and Green, E. (1996) *Taking Control or Juggling Jobs? An Analysis of the Employment Experiences of Women Workers who Leave Organisations to Become Entrepreneurs*, paper to the 14th Annual International Labour Process conference, Aston University.

Coleman, A. (1985) *Utopia on Trial*, Hilary Shipman.

Collins, R. (1990) Changing conceptions in the sociology of professions, in Torstendahl, R. and Burrare, M. (eds.) *The Formation of Professions: Knowledge State and Strategy*, Sage, London, pp. 11–23.

Comer, L. (1974) *Wedlocked Women*, Feminist Books, Leeds.

Commission for Racial Equality (1984) *Race and Council Housing in Hackney*, CRE, London.

Commission for Racial Equality (1985) *Race and Mortgage Lending*, CRE, London.

Commission for Racial Equality (1988) *Racial Discrimination in a London Estate Agency*, CRE, London.

Commission for Racial Equality (1989) *Racial Discrimination in Liverpool City Council*, CRE, London.

Commission for Racial Equality (1990) *Sorry it's Gone: Testing for Racial Discrimination in the Private Rented Sector*, CRE, London.

Connell, R.W. (1987) *Gender and Power: The Society, the Person and Sexual Politics*, Polity Press, Cambridge.

Consumers' Association see Taylor Nelson Consumer (1994).

Conway, J. (ed.) (1988) *Prescription for Poor Health: The Crisis For Homeless*

Families, The London Food Commission, London.

Cook, E. (1994) Talking shop in the men's room, *Independent*, September 25th, p. 22.

Cooper Marcus (1995) *House as a Mirror of Self: Exploring the Deeper Meaning of Home*, Conari, Berkeley.

Coote, A. and Campbell, B. (1982) *Sweet Freedom: the struggle for women's liberation* Pan Books, London.

Coyle, A. (1984) *Redundant Women*, The Women's Press, London.

Coyle, A. (1989) The limits of change: local government and equal opportunities for women, *Public Administration*, Vol. 67, Spring, pp. 39–50.

CRE see Commission for Racial Equality.

CRESR and SSRC (1993) *Barnsley City Challenge : Community Priorities and Attitudes*, Sheffield Hallam University.

Crime Concern (1993) *Planning the built environment: Women's Safety*, Briefing Paper no. 1, Crime Concern, Signal Point, Station Road, Swindon, Wilts.

Crow, G. (1989) The post-war development of the modern domestic ideal, in Allan, G. and Crow, G. (eds.) *Home and Family: Creating the Domestic Sphere*, Macmillan, London.

Csikszentmihalyi, M. and Rochberg-Halton, E. (1981) *The Meaning of Things: Domestic Symbols and the Self*, Cambridge University Press.

CSO see Central Statistical Office.

Dalla Costa, M. and James, S. (1972) *The Power of Women and the subversion of the community*, Falling Wall Press, Bristol.

Daly, M. (1979) *Gyn/Ecology: the metaethics of radical feminism*, The Woman's Press, London.

Darke, J. (1984), Women, architects and feminism, in Matrix, *Making Space: Women and the Man-Made Environment*, Pluto Press, London.

Darke, J. (1994) Women and the meaning of home, Chapter 2, pp. 11–30 in Gilroy, R. and Woods, R. (eds.) *Housing Women*, Routledge, London.

Darke, J., Conway, J. and Holman, C. with Buckley, K. (1992) *Homes for our Children*, National Housing Forum, London.

Darke, R. (1990) Introduction to Popular Planning, and A City Centre For People: Popular Planning in Sheffield, in Montgomery, J. and Thornley, A. (eds.), *Radical Planning Initiatives*, Gower, Aldershot.

Daunton, M.J. (1983) *House and home in the Victorian city: Working-class housing 1850–1914*, Edward Arnold, London.

Davidoff, L. and Hall, C. (1987) *Family Fortunes: Men and Women of the English Middle Class 1780–1850*, Hutchinson, London.

Davidson, M.J. and Cooper, D. (1992) *Shattering the Glass Ceiling, The Woman Manager*, Paul Chapman Publishing, London.

Davies, G. and Bell, J. (1991) The grocery shopper — is he different?

International Journal of Retail and Distribution Management, Vol. 19, no. 1, pp. 25–28.

Deem, R. (1986) *All Work and no play? the sociology of women's leisure*, Open University Press, Milton Keynes.

Dempsey, K. (1987) *Gender Inequality: the Exclusion and Exploitation of Women by Men in an Australian Rural Community*, paper to the First International Congress on the Future of Adult Life, April, 1987, Leewenhorst, The Netherlands.

Department of Employment (1975) *New Earnings Survey*, HMSO, London.

Department of Environment (1991) *Homelessness Code of Guidance for Local Authorities*, HMSO, London.

Department of the Environment (1992) *Development Plans, A Good Practice Guide*, HMSO, London.

Department of Environment (1993) *English House Condition Survey 1991*, HMSO, London.

Department of Environment (1994a) *Access to Local Authority and Housing Association Tenancies. A Consultation Paper*, HMSO, London.

Department of the Environment (1994) *Planning Out Crime*, Circular 5/94, Department of the Environment, London.

Department of Environment (1994c) *Community Involvement in Planning and Development Processes*, DoE Planning Research Programme, HMSO, London.

Department of the Environment (1995) Planning Policy Guidance Note 6, *Town Centres and Retail Developments*, HMSO, London.

Department of the Environment (annually and quarterly) *Housing and Construction Statistic (H&CS)*, London, HMSO.

Department of the Environment and Department of Transport (1993), *Reducing Transport Emissions through Planning*, London: HMSO.

Department of Health and Social Security (1988) *Low Income Families 1985*, HMSO, London.

Department of Social Security (1995) *Households Below Average Income: a Statistical Analysis, 1979–1992/93*, Government Statistical Service, HMSO, London.

Department of Transport (1993) *National Travel Survey 1989/91*, HMSO, London.

Department of Transport (1994) *National Travel Survey, 1991/93*, HMSO, London.

Department of Transport (ongoing) *COBA Manual*, HMSO.

Deutscher, I. (1968) The gatekeeper in public housing, in Deutscher, I. and Thompson, E. (eds.) *Among The People; Encounters With the Poor*, Basic Books, New York.

DHSS see Department of Health and Social Security.

Dobash, R.E. and Dobash, R. (1980) *Violence against wives: a case against the patriarchy*, Open Books, London.

Dobash, R.E. and Dobash, R. (1992) *Women, violence and social change*, Routledge, London.

Dodwell, C. (1995) *From staff hostels to foyers: an investigation of employment-related housing for young people*. Unpublished Masters Dissertation, School of Planning, Oxford Brookes University.

DOE see Department of Employment.

DoE see Department of the Enviroment.

Donzelot, J. (1979) *Policing the family*, Pantheon, New York.

Dorling, D. and Cornford, J. (1994) Market accentuates those negatives, in *ROOF* January-February pp. 12–13.

Dowling, R. (1993) Feminity, place and commodities: a retail case study, *Antipode*, Vol. 25, no. 4, pp. 295–319.

Drake, B. (1984) *Women in Trade Unions*, Virago, London.

DSS see Department of Social Security.

Duchene, C. and Pecheur, P. (1995) Women and decisions in the field of public transportation, Ch. 15, pp. 127–134 in Ottes, L., Poventud, E. *et al.* (eds.) op.cit.

Economist Intelligence Unit (1995) *Retail Trade Review*, No. 34, June.

Edwards, S. (1987) Provoking her own demise: from common assault to homicide, in Hanmer, J. and Maynard , M. (1987) op. cit.

Egerton, J. (1990) 'Out but not down: lesbians' experience of housing', *Feminist Review* 36, Autumn.

Ehrenreich, B. and English, D. (1979) *For her own good: 150 years of the experts' advice to women*, Pluto, London.

Eno, S. and Treanor, D. (1982) *The Collective Housing Handbook*, Lauriston Hall Publications, Castle Douglas.

Esam, P. and Berthoud, R. (1991) *Independent Benefits for Men and Women*, Policy Studies Institute, London.

Eurofem (1995) see Ministry of the Environment, Finland.

Euromonitor (1995) *Retail Trade International* 8th Edition, Vol. 2, Euromonitor, London.

European Commission (1994) *Employment in Europe 1994*, Directorate-General for Employment, Industrial Relations and Social Affairs, Office for Official Publications of the European Communities, Luxembourg.

Factory Inspectorate (1878–1974) *Reports of the Chief Inspector of Factories and Workshops to Her Majesty's Principal Secretary of State for the Home Department* (Annual Reports), HMSO, London.

Fava, S.F. (ed.) (1968) *Urbanism in World Perspective*, Thomas Y. Crowell Company, New York.

Fava, S.F. (1980) Women's place in the new suburbia, in Wekerle,

Peterson and Morley (eds.).

Featherstone, M. (1993) *Virtual Reality, cyberspace and the ageing body*, mimeo.

Financial Times (1995a) Department Stores are back in fashion, July 22.

Financial Times (1995b) Take your pick of Paris Markets, Nov. 25/26.

Finch, J. (1993) 'It's great to have someone to talk to' : ethics and politics of interviewing women, in Bell, C. and Roberts, H. (eds.) *Social Researching: politics, problems, practice*, Routledge & Kegan Paul, London, pp. 70–87.

Finch, J. and Summerfield, P. (1991) Social Reconstruction and the emergence of companionate marriage, in Clark, D. (ed.) *Marriage, domestic life and social change: writings for Jacqueline Burgoyne (1944–88)*, Routledge, London.

Fiske, J. (1989) Shopping for Pleasure, pp. 13–42 in *Reading the Popular*, Unwin Hyman, London.

Foster, P. (1983) *Access to Welfare: An Introduction to Welfare Rationing*, Macmillan, London.

Foucault, M. (1978) *The History of Sexuality*, Penguin, Harmondsworth.

Foulsham, J. (1990) Women's Needs and Planning — a Critical Evaluation of Recent Local Authority Practice, in Montgomery, J. and Thornley, A. (eds.) *Radical Planning Initiatives*, Gower, Aldershot.

Franck, K. (1994) Questioning the American dream: recent housing innovations in the United States, Chapter 11 in Gilroy and Woods (eds.).

Franck, K.A. and Ahrentzen, S. (eds.) (1991) *New Households New Housing*, Van Nostrand Reinhold, New York.

Franklin, A. (1990) Ethnography and housing studies, *Housing Studies*, Vol. 5, pp. 92–111.

French, M. (1978) *The Women's Room*, Sphere Books, London.

Frevert, U. (1989) *Women in German History: from Bourgeois Emancipation to Sexual Liberation*, Berg, Oxford and Providence, RI.

Gardiner, C. and Hill, R. (1996) Analysis of access to cars from the 1991 UK Census: samples of anonymised records: a case study of the elderly population of Sheffield, *Urban Studies*, Vol. 33, no. 2, pp. 269–281.

Gavron, H. (1968) *The Captive Wife*, Penguin, Harmondsworth.

Gershuny, J. (1983) *Social innovation and the division of labour*, Oxford University Press, Oxford.

Gershuny, J. (1993), Escorting children: impact on parental lifestyle, in Hillman, M. (ed.) *Children, Transport and the Quality of Life*, Ch. 8, pp. 62–75, Policy Studies Institute, London.

Giddens, A. (1990) *The Consequences of Modernity*, Stanford University Press, Stanford.

Gilroy, R. (1994) Women and owner occupation in Britain: first the prince, then the palace?, Chapter 3 in Gilroy and Woods (eds.).

Gilroy, R. and Woods, R. (eds.) (1994) *Housing Women*, Routledge, London.

Glendinning, C. and Millar, J. (eds.) (1987) *Women and Poverty in Britain*, Harvester Wheatsheaf, Hemel Hempstead.

Glendinning, C. and Millar, J. (eds.) (1992) *Women and Poverty in Britain: the 1990s*, Harvester Wheatsheaf, Hemel Hempstead.

GLC see Greater London Council.

Goffman, E. (1968) *Asylums*, Penguin, London.

Goodman, A. and Webb, S. (1995) *The Distribution of UK Household Expenditure 1979–92*, Institute for Fiscal Studies, London.

Goodwin, M. (1992) The changing local state, in Cloke, P. (ed.) *Policy and Change in Thatcher's Britain*, pp. 77–96, Pergamon, Oxford.

Goodwin, M. (1995) Poverty in the city: 'You can raise your voice but who is listening?', in Philo, C. (ed.) op. cit.

Gordon, D. and Forrest, R. (1995), *People and Places 2: Social and Economic Distinctions in England*, SAUS, Bristol.

Goss, S. (1984) Women's initiatives in local government, in Boddy, M. and Fudge, C. (eds.) *Local socialism? Labour Councils and New Left Alternatives*, Macmillan, London.

Goss, S., Stewart, L. and Wolmar, C. (1989) Making space — bringing feminism into the Town Hall, *Councils in Conflict: the Rise and Fall of the Municipal Left*, Macmillan, London.

Gough, J. (1994) Green light for super plan, *Glasgow Evening Times*, March 24th, p. 15.

Greater London Council (1986) *Changing Places*, GLC, London.

Greed, C. (1991) *Surveying Sisters: Women in a Traditional Male Profession*, Routledge, London.

Greed, C. (1994) *Women and Planning: Creating Gendered Realities*, Routledge, London.

Greed, C. (1996) Planning for women and other disenabled groups with reference to public toilet provision, *Planning and Environment A*, Vol. 28, pp. 573–588.

Green, E. (1996 forthcoming) Women and Leisure, in Kramarae, C. and Spender, D. *The Women's Studies Encyclopaedia*, Harvester Wheatsheaf, Hemel Hempstead.

Green, E. and Cohen, L. (1995) Women's Business: are women entrepreneurs breaking new ground, or simply balancing the demands of women's work in a new way? *Journal of Gender Studies*, Vol. 4, no. 3, pp. 297–314.

Green, E., Hebron, S. and Woodward, D. (1987a) *Gender and Leisure: a Study of Sheffield Women's Leisure*, Sports Council/ESRC, London.

Green, E., Hebron, S. and Woodward, D. (1987b) Women, Leisure and

Social Control, in Hanmer, J. and Maynard, M. (ed.).

Green, E., Hebron, S. and Woodward, D. (1990) *Women's Leisure, What Leisure?*, Macmillan, London.

Gregson, N. and Crewe, L. (1994) Beyond the high street and the mall: car boot fairs and the new geographies of consumption, *Area*, September, pp. 261–267.

Gregson, N. and Lowe, M. (1994) *Servicing the Middle Classes: Class, Gender and Waged Domestic Labour in Contemporary Britain*, Routledge, London.

Groves, D. (1992) Occupational pension provision and women's poverty in old age, in Glendinning, C. and Millar, J. (eds.) op. cit..

Gurney, C. (1995a) *Images and meanings of home and home ownership*, Paper given at conference on Gender Perspectives on Household Issues, University of Reading, April 1995.

Gurney, C. (1995b) '. . . *Oh, we wouldn't live in a council house*' Paper given at the British Sociological Association annual conference: Contested Cities, Social Processes and Spatial Forms, University of Leicester, April 1995.

Hakim, C. (1978) Sexual divisions within the labour force: occupational segregation, *Employment Gazette*, Vol. 86, no. 1, pp. 1264–1279.

Hakim, C. (1981) 'Job segregation: trends in the 1970s' *Social Policy*, Vol. 18, no. 4, pp. 471–503.

Hakim, C. (1989) 'Workforce restructuring, social insurance coverage and the black economy', *Journal of Social Policy*, Vol. 18, no. 4, pp. 471–503.

Hakim, C. (1995) 'Five feminist myths about women's employment', *British Journal of Sociology*, Vol. 46, no. 3, pp. 429–455.

Hall, A. (1974) *The Point of Entry*, Allen & Unwin, London.

Hall, S. (1994) 'New cultures for old', in Massey, D. and Jess, P. (eds.) *The Shape of the World*, Oxford University Press.

Halford, S. and Duncan, S. (1989) Implementing Feminist Policies in British Local Government, Working Paper 78, Centre for Urban and Regional Research, University of Sussex.

Hamilton, K. and Jenkins, L. (1989) 'Women and transport, in Grieco', Pickup and Whipp (eds.) (1989) *Gender, Transport and Employment*; republished in Roberts, *et al.* (1992) *Travel Sickness*, Ch. 6, pp. 57–74.

Hanmer, J. and Maynard, M. (eds.) (1987) *Women, Violence and Social Control*, Macmillan, London.

Hanmer, J. and Saunders, S. (1984) *Well Founded Fear*, Hutchinson, London.

Hanson, S. and Pratt, G. (1995) *Gender, Work and Space*, Routledge, London.

Haraway, D. (1994) A manifesto for cyborgs: science, technology and socialist feminism in the 1980s, in Seidman, S. (ed.) *The Postmodern Turn: New Perspectives on Social Theory*, Cambridge University Press.

Hargreaves, J. (1994) *Sporting Females, Critical Issues in the History and Sociology of Women's Sports*, Routledge, London.

Harker, L. (1996) *A Secure Future?: Social Security and the Family in a Changing World*, CPAG, London.

Hartmann, H. (1981) The unhappy marriage of Marxism and feminism: towards a more progressive union, in Sargent, L. (ed.) *The Unhappy Marriage of Marxism and Feminism: a debate on class and patriarchy*, Pluto Press, London.

Hayden, D. (1981) *The Grand Domestic Revolution*, MIT Press, Cambridge.

Healey, P. (1989) *Planning for the 1990s*, Working Paper, series No. 7, Department of Town and Country Planning, University of Newcastle-upon-Tyne.

Healey, P. (1990) Places, people and policies, *Town and Country Planning*, January 1990.

Healey, P. (1994) *Bringing Women into Urban and Regional Planning: Slow Progress to Big Gains*, paper to the Council of Europe Colloquy: The challenges facing European Society with the approach of the year 2000: role and representation of Women: urban and regional planning aiming at sustainable development, Ornskolvick, Sweden, 24–26 March.

Hearn, J. (1992) *Men in the public eye*, Routledge, London.

Heath, S. and Dale, A. (1994) Household and family formation in Great Britain: the ethnic dimension, *Population Trends 77*, autumn, pp. 5–13.

Henderson, J. and Karn, V. (1987) *Race, Class and State Housing*, Gower, Aldershot.

Henderson, K.A. and Bialeschi, M.D. (1992) Leisure Research and the Social Structure of Feminism, *Society and Leisure*, Vol. 15, no. 1, pp. 63–75.

HMSO see Department of Employment, Department of the Environment, Home Office, House of Commons Environment Committee, Ministry of Transport and Ministry of Housing and Local Government, OPCS.

Herrin, J. (1983) In search of Byzantine women, in Cameron and Kuhrt (eds.).

Hewitt, P. (1975) *Rights for women: a guide to the Sex Discrimination Act, Equal Pay Act, paid maternity leave, pension schemes and unfair dismissal*, NCCL, London.

Hey, V. (1986) *Patriarchy and Pub Culture*, Tavistock, London.

Higgins, M. and Davies, L. (1996) Planning for women — how much has been achieved? *Planning Practice and Research*, Vol. 22, no. 1.

Hill, O. (1910) letter to *The Times*, July, quoted in Moberley Bell, E. (1942) *Octavia Hill*, Constable, London.

Hill, R.D. (1986) Urban transport: from technical process to social policy, in Lawless, P. and Raban, C. (1986) *The Contemporary British City*, Paul Chapman, Chapter 5, pp. 85–106.

Hill, R.D. (1993) Planning in transport? the role of land use planning in transport provision in the metropolitan Areas of England, *Journal of Transport Geography*, Vol. 1, no. 2, pp. 131–134.

Hillier, B. and Hanson, J. (1984) *The Social Logic of Space*, Cambridge University Press.

Hillman, M. (ed.) (1993) *Children, Transport and the Quality of Life*, Policy Studies Institute, London.

Holland, J. (ed.) (1984) *Feminist Action*, Battle Axe Books, Middlesex.

Holmans, A.E. (1987) *Housing Policy in Britain*, Croom Helm, London.

Home Office (1984) *Joint Circular on Crime Prevention*, HMSO, London.

Home Office (1989) *Tackling Crime*, HMSO, London.

Hood, M. and Woods, R. (1994) 'Women and Participation', in Gilroy, R. and Woods, R. (ed.), *Housing Women*, Routledge, London.

hooks, bell (1991) Chapter in *Yearning: Race, Gender and Cultural Politics*, Turnaround, London.

Horrell, S. (1994) Household time allocation and women's labour force participation, Chapter 6 in Anderson, M., Bechhofer, F. and Gershuny, J. (eds.) *The Social and Political Economy of the Household*, Oxford University Press, Oxford.

Horelli, L. and Vespa, K. (1994) In Search for Supportive Structures for Everyday Life in Altman, I. and Churchman, A. (eds.) *Women and the Environment*, Plenum, New York.

House of Commons Environment Committee (1994) *Fourth Report: Shopping Centres and their Future*, HMSO, London.

Hudson, R., Schech, S. and Hansen, K. (1992) *Jobs for the Girls? The New Private Sector Economy of Derwentside District*, Durham University, Geography Department, Occasional Publications.

Hufton, O. (1995) *The Prospect Before Her: a History of Women in Western Europe*, Harper Collins, London.

Hughes, D. (1989) Paper and people: the work of the casualty reception clerk, *Sociology of Health and Illness*, 11 (4), pp. 382–408.

Hunt, A. (1968) *A Survey of Women's Employment for the Ministry of Labour*, HMSO, London.

Hunt, G. and Saterlee, S. (1987) Darts, drink and the pub: the culture of female drinking. *Sociological Review*, August 1987.

Hunt, P. (1980) *Gender and Class Consciousness*, Macmillan, London.

Hunter, P. L. and Whitson, D. (1991) Women, Leisure and Familism, *Leisure Studies*, Vol. 10, pp. 219–233.

Huxley, M. 1988 'Feminist urban theory: gender, class and the built environment', *Transition*, Winter, pp. 39–43.

The Independent on Sunday (1994) Customer relations: Toddlers drawn into stores war, March 13th, p. 15.

Jacobs, J. (1967) *The Death and Life of Great American Cities*, New York, Random House, Penguin, Harmondsworth (1964).

Jackson, P. and Holbrook, B. (1995) Multiple Meanings: shopping and the cultural politics of identity, *Environment and Planning A.*, Vol. 27, no. 12, pp. 1913–30.

Jackson, P. and Thrift, N. (1995) Geographies of consumption, in Miller, D. (ed.) *Acknowledging Consumption*, Routledge, London.

Jeffers, S. and Hoggett, P. (1995) Like counting deckchairs on the Titanic: a study of institutional racism and housing allocations in Haringey and Lambeth, *Housing Studies*, Vol. 10 (3), pp. 325–344.

Kandola, R. and Fullerton, J. (1994) Diversity: More than Just an Empty Slogan, *Personnel Management*, November.

Kelly, L. (1988) *Surviving Sexual Violence*, Polity, Cambridge.

Kempson, E., Bryson, A., Rowlingson, K. (1994) *Hard Times? How Poor Families Make Ends Meet*, PSI, London.

Kennedy, H. (1992) *Eve was Framed: Women and British Justice*, Chatto and Windus, London.

Kirby, D. (1993) Working Conditions and the trading week, in Bromley, R. and Thomas, C. (eds.) (1993) *Retail Change, Contemporary Issues*, UCL, London.

Kureshi, H. (1991) *The Buddha of Suburbia*, Faber, London.

Land, H. (1976) Women: supporters or supported?, in Barker, D. L. and Allen, S. (eds.) *Sexual Divisions and Society: Process and Change*, Tavistock Publications, London.

Land, H. (1991) 'Time to care', in Maclean, M. and Groves, D. *Women's Issues in Social Policy*, Routledge, London.

Lansley, S., Goss, S. and Woolmar, C. (1989) *Councils in Conflict: The Rise and Fall of the Municipal Left*, Macmillan, London.

Lefkowitz, M. (1983) Influential women, in Cameron and Kuhrt (eds.).

Leicester City Council, Women's Equality Unit (1993a) *Women and Safety Report*, Leicester City Council.

Leicester City Council, Women's Equality Unit (1993b) *Women and Safety Conference Report*, Leicester City Council.

Lessing, D. (1973) *The Golden Notebook*, Panther Books, London.

Lewis, J. (1992) *Women in Britain since 1945*, Basil Blackwell, Oxford.

Lewis, J. and Foord, J. (1984) New Towns and gender relations in old industrial regions: women's employment in Peterlee and East Kilbride, *Built Environment*, Vol. 10, no. 1, pp. 42–52.

Lewis, J. and Piachaud, D. (1992), Women and Poverty in the Twentieth

Century, in Glendinning, C. and Millar, J. (eds.) op. cit..

Lewis, J. (1993) (ed.) *Women and Social Policies in Europe: work, family and the state*, Edward Elgar, Aldershot.

Liddington, J. and Norris, J. (1978) *One Hand Tied Behind Us: the Rise of the Women's Suffrage Movement*, Virago, London.

Lidstone, P. (1994) Rationing housing to the homeless applicant, *Housing Studies* 9(4) pp. 459–472.

Lipsky, M. (1980) *Street-Level Bureaucracy*, Russell Sage, New York.

Lister, R. (1989) Assessment of the Fowler Review in Dilnot, A. and Walker, I. (eds.) *The Economics of Social Security*, Oxford University Press, Oxford.

Lister, R. (1992) *Women's Economic Dependency and Social Security*, Equal Opportunities Commission, Manchester.

Little, J. (1994) *Gender, Planning and the Policy Process*, Pergamon Press, Oxford.

Little J. Peake L. and Richardson P. (1988) *Women in Cities: Gender and the Urban Environment*, Macmillan, Basingstoke.

Little, J. (1994) Women's initiatives in town planning in England: a critical review, in *Town Planning Review*, Vol. 65, no.3, pp. 261–276.

Little, J., Peake, L. and Richardson, P. (1988) Introduction: geography and gender in the Urban Environment, in Little, J., Peake L. and Richardson, P. (eds.) *Women in Cities*, Macmillan, London.

Lo,Lucia (1994) Exploring teenage shoplifting behaviour — a choice and constraint approach, *Environment and Behaviour*, September, Vol. 26, no. 5, pp. 613–639.

Local Transport Today see Starkie.

MacEwen Scott, A. (1994) Gender segregation in the retail industry, in MacEwen Scott, A. *Gender Segregation and Social Change*, Oxford University Press, Oxford.

MacGregor, S. and Pimlott, B. (eds.) (1991) *Tackling the Inner Cities: the 1980s Reviews, Prospects for the 1990s*, Clarendon Press, Oxford.

Mackie, L. and Pattullo, P. (1977) *Women at Work*, Tavistock Publications, London.

Maddock, S. and Parkin, D. (1994) Gender cultures: how they affect men and women at work, in Davidson, M. J. and Burke, R. J. (eds.) *Women in Management: Current Research Issues*, Paul Chapman Publishing, London.

Maddock, S. and Parkin, D. (1996) Gender Cultures: Women's Choices and Strategies at Work, in Bilsbury, J. (ed.) *The Effective Manager*, Sage, London, pp. 101–112.

Madigan, R. and Munro, M. (1991) Gender, House and "Home": social meanings and domestic architecture in Britain, *Journal of Architectural and Planning Research* Vol. 8, no. 2, pp. 116–131.

Marcus, S. (1971), Planners — who are you? *Journal of the Royal Town Planning Institute*, Vol. 57, no. 2.

Marks and Spencers (1995) *Report and Accounts for 1995*, M&S, London.

Martin, J. and Roberts, C. (1984) *Women and Employment: A Lifetime Perspective*, HMSO, London.

Martin, R. and Wallace, J. (1984) *Working Women in Recession: Employment, Redundancy and Unemployment*, Oxford University Press.

Mason, J. (1988) No peace for the wicked: older married women and leisure, in Talbot, M. and Wimbush, E. (eds.) *Relative Freedoms*, Open University Press, Milton Keynes.

Mason, J. (1989) Reconstructing the public and the private: the home and marriage in later life, Chapter 7, in Allan, G. and Crow, G. (eds.) *Home and Family: Recreating the domestic sphere*, Macmillan, Basingstoke.

Massey, D. (1994) *Space, Place and Gender*, Polity Press, Cambridge.

Matrix Group (1984) *Making Space: Women and the Man-Made Environment*, Pluto Press, London.

Maynard, M. and Purvis, J. (eds.) (1994) *Researching Women's Lives from a Feminist Perspective*, Taylor & Francis, London.

McCamant, K. M. and Durrett, C. R. (1991) Co-housing in Denmark, Chapter 5 in Franck and Ahrentzen (eds).

McDougall, M. (1995), *Turning the Glass Ceiling into a Window of Opportunity*, IPD West of Scotland Branch Meeting: 16 November.

McFarlane, B. (1984) Homes fit for heroines, in Matrix (ed.) *Making Space: momen and the Man-made Environment*, Pluto, London.

McKee, L. and Bell, C. (1985) His unemployment, her problem, in Allen, S. *et al.* (eds.), *The Experience of Unemployment*, Macmillan, London.

McRae, S. and Daniel, W.W. (1991) *Maternity Rights: the Experience of Women and Employers: First Findings*, Policy Studies Institute, London.

McRobbie, A. (1993) Shut up and dance: youth culture and changing modes of feminity, *Cultural Studies*, Vol. 7, pp. 406–426.

Merrett, S. (1979) *State Housing in Britain*, Routledge & Kegan Paul, London.

Millar, J. (1989) Social Security, equality and women in the UK, *Policy and Politics*, Vol. 17, no. 4, pp. 311–19.

Miller, M. (1990) *Bed and Breakfast: Women and Homelessness Today*, The Women's Press, London.

Millet, K. (1977) *Sexual Politics*, Virago, London.

Ministry of Health (1949) *Housing Manual 1949*, HMSO, London.

Ministry of Housing and Local Government (1952) *Houses 1952*, HMSO, London.

Ministry of Housing and Local Government (1968) *House Planning: a Guide to User Needs* (with a check list), Design Bulletin 14, HMSO, London.

Ministry of the Environment, Finland (1995) *Eurofem Gender and the Human Environment*, Proceedings from the Second Working Meeting, at the Ministry of the Environment, The Hague, The Netherlands.

Ministry of Transport and Ministry of Housing and Local Government (1964), Buchanan Report on Traffic in Towns, Circular 1/64, HMSO, London.

Mitchell, J. (1971) *Woman's Estate*, Penguin, Harmondsworth.

Mooney, J. (1995) *Violence, Space and Gender: the Social and Spatial Parameters of Violence Against Women and Men*, paper presented at the British Sociological Association Annual Conference, University of Leicester.

Morgan, J. (1991) *Safer Communities: The Local Delivery of Crime Prevention Through a Partnership Approach*, Report of the Standing Conference on Crime Prevention, Home Office, London.

Morgan, P. (1995) *Farewell to the Family? Public Policy and Family Breakdown in Britain and the USA*, IEA, London.

Morphet, J. (1992) Women local authority Chief Executives: roots and routes, *Local Government Policy Making*, Vol. 19, no. 3, December 1992.

Morris M. J. (1991–2) ' "Us" and "them"? Feminist research, community care and disability', *Critical Social Policy* 33, pp. 22–39.

Morris, L. (1987) Constraints on gender, *Work, Employment and Society*, Vol. 1, pp. 85–106.

Morris, L. (1990) *The Workings of the Household*, Polity Press, Cambridge.

Morris, L. (1993) Domestic labour and the employment status of married couples, *Capital and Class*, Vol. 49, pp. 37–52.

Morris, L. and Llewellyn, T. (1991) *Social Security Provision for the Unemployed: Report to the Social Security Advisory Committee*, HMSO, London.

Moylan, S., Millar, J. and Davies, R. (1984) *For Richer, For Poorer, DHSS Cohort Study of Unemployed Men*, DHSS, Research Report No 11, HMSO, London.

Mumford, L. (1945) On the future of London, *Architectural Review*, Vol. xcvii , no. 577, January.

Mumford, L. (1961) *The City in History*, Penguin, Harmondsworth.

Muthesius, S. (1982) *The English Terraced House*, Yale University Press, New Haven.

Nadin, V. and Jones, S. (1990) A profile of the profession, *The Planner*, 26 January.

National Association of Local Women's Committees (1986) *Responding with Authority: Local Authority Initiatives to Counter Violence Against Women*, Pankhurst Press, London.

Newman, J. (1993) Women, Management and Change, *Local Government Policy Making*, No. 20, No. 2 October.

200   CHANGING PLACES

Newman, O. (1973) *Defensible Space: People and Design in the Violent City*, The Architectural Press, London.
Nielsen, J.M. (1990) (ed.) *Feminist Research Methods: exemplary readings in the social sciences*, Westview Press, London.

Oakley, A. (1974) *The Sociology of Housework*, Martin Robertson, London.
Oakley, A. (1981) Interviewing women: a contradiction in terms, in Roberts, H. (ed.) *Doing Feminist Research*, Routledge & Kegan Paul, London.
OECD (Organisation for Economic Co-operation and Development) (1994) *Labour Force Statistics 1972–1992*, Geneva.
OECD (Organisation for Economic Co-operation and Development) (1994), *High Level Conference 'Women in the City: Housing, Services and the Urban Environment'*, Conclusions of the Chair, Paris, France.
OECD (Organisation for Economic Co-operation and Development), European Conference of Ministers of Transport (1991) *Structural Changes in Population and Impact on Passenger Transport*: Report of 88th Round Table on Transport, Paris 13th–14th June.
OPCS (Office of Population Censuses and Surveys) (1980) *OPCS Monitor*, HMSO, London.
OPCS (Office of Population Censuses and Surveys) (1993) *Social Trends*, OPCS, HMSO, London.
Office of Population Censuses and Surveys, (1994) *General Household Survey 1992*, HMSO, London.
OPCS (Office of Population Censuses and Surveys) (1995a) *Social Trends 25*, HMSO, London.
OPCS (Office of Population Censuses and Surveys) (1995b) *General Household Survey 1993*, HMSO, London.
OPCS (Office of Population Censuses and Surveys) (various dates) *Census*, HMSO, London.
Oppenheim, C. (1993) *Poverty: the facts*, CPAG, London.
Oppenheim, C. and Harker, L. (1996) *Poverty: The Facts*, Third edition, CPAG, London.
Ottes, L., Poventund, E., van Schendelen, M. and Segond von Banchet, G. (eds.) (1995) *Gender and the Built Environment Emancipation in Planning, Housing and Mobility in Europe*, Van Goreum, Assen, Netherlands.

Page, D. (1994) *Developing Communities*, Sutton Hastoe Housing Association, Teddington.
Pahl, J. (1983) The Allocation of Money and the Structuring of Inequality within Marriage, *Sociological Review*, 31, pp. 237–62.
Pahl, R.E. (1984) *Divisions of Labour*, Blackwell, Oxford.
Pahl, J. (1984) The allocation of money within the household, in

Freeman, M. (ed.) *The State, The Law and the Family*, Tavistock, London.

Parker, R. (1975) Social administration and scarcity, in Butterworth, E. and Holman, R. (eds.) *Social Welfare in Modern Britain*, Fontana, Glasgow.

Pateman, C. (1988) *The Sexual Contract*, Polity Press, Cambridge.

Payne, S. (1991) *Women, Health and Poverty: an introduction*, Harvester Wheatsheaf, Hemel Hempstead.

Penn, R. and Wirth, B. (1993) Employment patterns in contemporary retailing: gender and work in five supermarkets, *The Service Industries Journal*, Vol. 13, no. 4, pp. 252–266.

Philo, C. (ed.) (1995) *Off the Map: the Social Geography of Poverty in the UK*, CPAG, London.

Pickup, L. (1984) Women's gender role and its influence on their travel behaviour, *Built Environment*, Vol. 10, pp. 61–68.

Pickup, L. (1988) Hard to get around: a study of women's travel mobility, in Little *et al.* (eds.) op. cit.

Pickup, L. *et al.* (1991) *Bus Deregulation in the Metropolitan Areas*, Avebury.

Pillinger, J. (1992) *Feminising the Market: Women's Pay and Employment in the European Community*, Macmillan, London.

Pinch, S. and Storey, A. (1992) Who does what, where?: a household survey of the division of domestic labour in Southampton, *Area*, Vol. 24, no., pp. 5–12.

Pinchbeck, I. (1981) *Women Workers and the Industrial Revolution 1750–1850*, Virago, London.

Prottas, J, (1979) *People-Processing: The Street-level Bureaucrat in Public Service Bureaucracies*, Lexington Books, Lexington MA.

Pugh, C. (1990) A New Approch to Housing Theory: Sex, Gender and Domestic Economy, *Housing Studies*, April 1990 Vol. 5, no. 2. pp. 112–29..

Qureshi, H. and Walker, A. (1989) *The Caring Relationship*, Macmillan, Basingstoke.

RAC Foundation for Motoring and the Environment (1995), *Car Dependence*, Royal Automobile Club, London.

Radford, J (1987) Policing male violence, in Hanmer, J. and Maynard, M. (eds.) op.cit.

Ravetz, A. (1989) A view from the interior, in Attfield, J. and Kirkham, P. (eds.) *A View from the Interior*, The Women's Press, London.

Reekie, G. (1993) *Temptations: Sex Selling and the Department Store*, Allen and Unwin, Sydney.

Rees, T. (1992) *Women and the Labour Market*, Routledge, London.

Reeves, D. (1995) Developing effective public consultation: a review of

Sheffield's UDP process, *Planning Practice and Research*, Vol. 10, No. 2

Rein, M. and Erie, S. (1988) Women and the welfare state, in Mueller, C.M. (ed.) *The Politics of the Gender Gap*, Sage, London.

Rendall, J. (1990) *Women in an Industrialising Society: England 1750–1880*, Basil Blackwell, Oxford.

Rendel, M. (1978) Legislating for equal pay and opportunity for women in Britain, *Signs*, Vol. 3, no. 4, pp. 897–908.

Rich, A. (1980) *On Lies, Secrets and Silence: Selected Prose 1966–1978*, Virago, London.

Riseborough, M. and White, J. (1994) *Women and Safety: New Wine in Old Bottles?* paper to the International Sociological Association Conference, Bielefield, Germany, July 1994.

Ritchie, J. (1990) *Thirty Families: their Living Standards in Unemployment*, DSS Research Report No. 1, HMSO, London.

Roberts, H. (1981) (ed.) *Doing Feminist Research*, Routledge & Kegan Paul, London.

Roberts, M. (1991) *Living in a Man-Made World*, Routledge, London.

Roberts, N. (1992) *Whores in History: Prostitution in Western Society*, HarperCollins, London.

Robertson, E. (1984) Single person lifestyle and peripheral estate residence: a geographic investigation in Drumchapel, *Town Planning Review*, Vol. 55, no. 2, pp. 197–213.

Robins, G. (1983) The god's wife of Amun in the 18th dynasty in Egypt, in Cameron and Kuhrt (eds.).

Rojek, C. (1994) Leisure and the Dreamworld of Modernity, in Henry, I. (ed.), *Leisure; Modernity, Postmodernity and Lifestyles*, Leisure Studies Association, No. 48, Brighton.

ROOF Briefing (1995) no. 12, October.

Rosin, H. and Korabik, K. (1992) Corporate flight of women managers, *Women in Management Review*, Vol. 7, no. 3, pp. 31–35.

Ross, K. (1992) Inequality Circles, *Local Government Policy Making*, Vol. 19, no. 3, December.

Ross, S. (1991), Planning and the Public Interest, TCPSS Proceedings, 1991, TCPSS Prize Paper, *The Planner*, 13th December.

Rosser, K.C. and Harris, C.C. (1965) *The Family and Social Change*, Routledge & Kegan Paul, London.

Rowbotham, S. (1973) *Woman's Consciousness, Man's World*, Penguin, Harmondsworth.

Rowntree B.S. (1901) *Poverty: a Study of Town Life*, Macmillan, London.

Royal Town Planning Institute (1988) *Planning for Shopping in the 21st century, the Report of the Retail Planning Working Party*, Royal Town Planning Institute, London.

Royal Town Planning Institute (1989) *Planning for Choice and Opportunity*, Royal Town Planning Institute, London.

Royal Town Planning Institute (1995) *Planning for Women*, Practice Advice Note No. 12, Royal Town Planning Institute.

RTPI see Royal Town Planning Institute.

Rural Development Commission (1994) *Rural Services: Challenges and Opportunities*, RDC, London.

Ryan, J. (1994) Women, Modernity and the City, *Theory, Culture and Society*, Vol., 11, pp. 35–63.

Rybczinski, W. (1986) *Home: a Short History of an Idea*, Viking, New York.

Rydin, Y. (1993) *The British Planning System: an Introduction*, Macmillan, Basingstoke.

Saegert, S. and Winkel, G. (1980) The home: a critical problem for changing sex roles, in Wekerle, Peterson and Morley, op. cit.

Safe Estates for Women Group (1992) *Safe Estates for Women*, Bloomsbury Safety Audit, Safe Estates for Women's Group.

Sandercock, L. and Forsyth, A. (1992) A gender agenda: new directions for planning theory, *American Planning Association Journal*, Vol. 58, no. 1, Winter 1992, pp. 49–59.

Sanderson, M. (1987) *Educational Opportunity and Social Change in England*, Faber & Faber, London.

Saraceno, C. (1991) Changing women's life course patterns in Italy: gender, cohort and social class differences, in Heinz, W.R. (ed.) *The Life Course and Social Change: Comparative Perspectives*, Deutscher Studien Verlag, Weinheim, Germany.

Saunders, P. (1990) *A Nation of Home Owners*, Unwin Hyman, London.

Scott, H. (1984) *Working Your Way to the Bottom: the feminisation of poverty*, Pandora Press, London.

Scourfield, J. (1995) *Changing Men: UK Agencies working with men who are violent towards their women partners*, Social Work Monographs, University of East Anglia.

Scully, D. (1990) *Understanding Sexual Violence: a study of convicted rapists*, Unwin Hyman, London.

Sharpe, S. (1984) *Double Identity, The Lives of Working Mothers*, Penguin, London.

Sheffield City Council, Department of Land and Planning (1987) *Central Area District Plan: The Hearing for Advisory Groups*, unpublished.

Sheffield City Council (1992), Department of Land and Planning (1992), *Report to Positive Action Sub-Committee*, Women's Transport and Development Forum, 24th August, 1992.

Sheffield City Council (1993) *Consultation Report on the Unitary Development Plan*, Sheffield City Council Planning Department, Sheffield.

Sheffield City Council (1993), Department of Land and Planning (1993), *Discussion Paper on Women's Transport and Development Forum*, unpublished.

Shelter (1995) *No Place To Learn*, Shelter, London.

Signs (1980) special issue on Women and the American City.

Sly, F. (1994a) Ethnic groups in the labour market, *Employment Gazette*, May 1994, pp. 147–158, HMSO, London.

Sly, F. (1994b) Mothers in the labour market, *Employment Gazette*, November 1994, pp. 403–412, London.

Smith, G. (1991) Grocery shopping patterns of the ambulatory urban elderly, *Environment and Behaviour*, Vol. 23, no. 1, pp. 86–114.

Smith, S. J. (1989) *The Politics of 'Race' and Residence*, Polity, Cambridge.

Snell, M. (1986) Equal pay and sex discrimination, in Feminist Review (eds.) *Waged Work: a Reader*, Virago Press, London.

Social Trends (1995) *Social Trends 25*, HMSO, London.

Sommer, R., Wynes, M. and Brinkley, G. (1992) 'Social facilitation effects in shopping behaviour', *Environment and Behaviour*, Vol. 24, no. 3, pp. 285–297.

Spence, A. and Podmore, D. (1987, eds.) *In a Man's World, Essays on Women in Male Dominated Professions*, Tavistock, London.

Spring-Rice, M. (1939) *Working Class Wives*, Penguin, Harmondsworth.

Starkie D. (1995) Letter, *Local Transport Today*, 12 October, 1995, p. 15.

Stanko, E.A. (1985) *Intimate Intrusions*, Unwin Hyman, London.

Stanko, E.A. (1987) Typical violence, normal precaution: men, women and interpersonal violence in England, Wales, Scotland and the USA, in Hanmer, J. and Maynard, M. op.cit.

Stanko, E.A. (1988) 'Hidden violence against women', in Maguire, M. and Pointing, J. (eds.) *Victims of Crime: A New Deal?* Open University Press, Milton Keynes.

Stanko, E.A. (1990) *Everyday Violence: How women and men experience physical and sexual danger*, Pandora Press, London.

Stanko, E.A. (1993) Ordinary fear: women, violence and personal safety, in Bart, P.B. and Moran, E.G. op.cit.

Stanko, E.A. (1994) Men's individual violence, in Newburn, T. and Stanko, E.A. (eds.) (1994) *Men, Masculinities and Crime: Just Boys Doing Business?*, Routledge, London.

Stanley, L. (1988) Historical sources for studying work and leisure in women's lives, in Wimbush *et al.*

Stanley, L. (1990) (ed.) *Feminist Praxis: Research, Theory and Epistemology in Feminist Sociology*, Routledge, London.

Stokes, G. and Taylor (1995) in Pickup, R, Jowell, J., Curtice, A., Park, L., Brook and Ahrendt, D. eds. ,1995, *British Social Attitudes*, the 12th Report of Social and Community Planning Research, Dartmouth, Aldershot.

Summerfield, P. (1984) *Women Workers in the Second World War: Production and Patriarchy in Conflict*, Croom Helm, London.

Swenarton, M. (1981) *Homes Fit for Heroes*, Heinemann, London.

Tawney, R.H. (1931) *Equality*, George Allen and Unwin, London.

Taylor, B. (1982), Planning: the need to plan for women, *Planning* 479, 30 July.

Taylor Nelson Consumer (1994) *Omnimas Shopping Survey* prepared for the Consumers' Association, Taylor Nelson Consumer, Epsom.

Thomas, A. and Niner, P. (1989) *Living in Temporary Accommodation: A Survey of Homeless People*, HMSO, London.

Thornley, A. (1989) *Whatever Happened to Participation in Planning? The British Experience Under Thatcherism*, paper presented to the 3rd Annual Congress of the Association of European Planning Schools, Tours, France.

Tivers, J. (1988) Women with young children: constraints on activities in the urban environment, in Little *et al.* (eds.).

Tomalin, C. (1992, revised edn.) *The Life and Death of Mary Wollstonecraft*, Penguin, Harmondsworth.

Tong, R. (1992) *Feminist Thought: a comprehensive introduction*, Routledge, London.

Town and County Planning Association (1986) *Planning for Women — An Evaluation of Consultation in Three London Boroughs*, London Planning Aid Service, Research Paper 2, Town and Country Planning Association.

Townsend, P. (1987) *Poverty and Labour in London*, Child Poverty Action Group, London.

Trench, S., Oc, T. and Tiesdell, S. (1992) Safer Cities for Women: Perceived risks and planning measures, *Town Planning Review*, Vol. 63, no. 3, pp. 276–296.

URBED (1994) *Vital and Viable Town Centres: Meeting the Challenge*, HMSO, London.

Urry, J. (1995) *Consuming Places*, Routledge, London.

Valentine, G. (1990) Women's fear and the design of public space, *Built Environment*, Vol. 16, no. 4, pp. 288–303.

Valentine, G. (1992) London's streets of fear, in Thomley, A. (ed.) *The Crisis of London*, Routledge, London.

Valentine, G. (1995) Out and about: geographies of lesbian landscapes, *International Journal of Urban and Regional Research*, Vol. 19, no. 1, pp. 96–111.

Venn, S. (1985) *Singled Out: Local Authority Policies for Single People*, CHAR, London.

Viner, K. (1994) Girls just wanna have fun, the *Guardian*, December 15, 1994.

Vogler, C. (1989) *Labour Market Change and Patterns of Financial Allocation Within Households*, Working Paper no. 12, ESRC, Social Change and Economic Life Initiative, Oxford.

Walby, S. (1986) *Patriarchy at Work*, Polity, Cambridge.

Walby, S. (1990) *Theorizing Patriarchy*, Basil Blackwell, Oxford.

Walby, S. (1990) Women's employment and the historical periodisation of labour, in Corr, H. and Jamieson, L. *The Politics of Everyday Life*, Macmillan, London.

Walker, A. (1992) The Poor Relation: poverty among older women in Glendinning, C. and Millar, J. (eds.) op. cit.

Walker, C. (1993) *Managing Poverty: the limits of social assistance*, Routledge, London.

Walker, S. (1983) Women and housing in classical Greece: the archaeological evidence, in Cameron and Kuhrt (eds.).

Ware, V. (1992) *Beyond the Pale: White Women, Racism and History*, Verso, London.

WEB (undated) *Shopping: I shop therefore I am*, Women and the Built Environment, London.

Watson, S. with Austerberry, H. (1986) *Housing and Homelessness: a Feminist Perspective*, Routledge & Kegan Paul, London.

Watson, S. and Gibson, K. (eds.) (1994) *Post Modern Cities and Spaces*, Blackwell, Oxford.

WDS see Women's Design Service.

Wearing, B. and Wearing, S. (1988) All in a day's leisure: gender and the concept of leisure, *Leisure Studies*, Vol. 7, pp. 111–123.

Webb, S. (1993) Women's Incomes: past, present and prospects, *Fiscal Studies*, Vol. 14, part 4, Institute for Fiscal Studies.

Wekerle, G., Peterson, R. and Morley, D. (eds.) (1980) *New Space for Women*, Westview Press, Boulder, Colorado.

Wekerle, G. and Whitzman, C. (1995) *Safe Cities, Guidelines for Planning, Design and Management*, Van Nostrand Reinhold, New York.

Welsh Women's Aid (1986) *The Answer is Maybe . . and That's Final*,WWA, Cardiff.

Westlake, T. (1993) Disadvantaged Consumers, in Bromley, R. and Thomas, C. (eds.) (1993) *Retail Change, Contemporary Issues*, UCL, London.

Willmott, P. and Hutchison, R. (1992) *Urban Trends 1: a Report on Britain's Deprived Urban Areas*, Policy Studies Institute, London.

Willmott, P. and Murie, A. (1988) *Polarisation and Social Housing*, Policy Studies Institute, London.

Wilson, E. (1991) *The Sphinx in the City*, Virago, London.

Wimbush, E. (1986) Women, Leisure and Well-Being: Final report, Edinburgh: Centre for Leisure Research, Dunfermline College of

Physical Education.

Wimbush, E. and Talbot, M. (1988) *Relative Freedoms: Women and Leisure*, Open University Press, Milton Keynes.

Wirth, L. (1968, originally published 1938) Urbanism as a way of life, in Fava (ed.)

Woodward A. (1991) 'Communal housing in Sweden: a remedy for the stress of everyday life?', Chapter 4 in Franck and Ahrentzen (eds.).

Woodward, D., Green, E. and Hebron, S. (1989) Bouncing the balls around or keeping them in the air? The sociology of women's leisure and physical recreation, in Parker, S. (ed.) *Leisure, Labour and Lifestyles: International Comparisons*, Vol. 2, Leisure Studies Association, pp. 76–91.

Women's Design Service (1992) *Unitary Development Plans*, July, WDS, London.

Women's Design Service (1994) *Are Town Centres Managing?*, Broadsheet No. 12, London Women and Planning Forum, WDS, London.

Women's Design Service (undated) *Women's Safety on Housing Estates*. WDS, London.

Woolf, M. (1994) Shoppers switch to cars, *Independent*, August 11, p. 4.

Yeandle, S. (1982) Variation and flexibility: key characteristics of female labour, *Sociology*, Vol. 16, no. 3, pp. 422–430.

Yeandle, S. (1984) *Women's Working Lives: Patterns and Strategies*, Tavistock, London.

Yeandle, S. (1987) Married Women at Midlife: Past Experience and Present Change, in Allatt, P., Keil, T., Bryman, A. and Bytheway, B. *Women and the Life Cycle: transitions and turning-points*, Macmillan, Basingstoke.

Yeandle, S. (1993) *Women of Courage: 100 years of women factory inspectors*, HMSO, London.

Yeandle, S. and Morrell, H. (1994) An Evaluation of Derby Safer Cities Women's Safety Strategy, CRESR, Sheffield Hallam University and Derby Safer Cities (unpublished).

Young, J. (1988) Risk and Fear of Crime: The Politics of Victimisation Surveys, in Maguire, M. and Pointing, J. (eds.) *Victims of Crime: A New Deal?* Sage, London.

Young, J. (1992) Ten Points of Realism, in Young, J. and Matthews, S. (eds.) *Rethinking Criminology, The Realist Debate*, Sage, London.

Young M. and Willmott, P. (1957) *Family and Kinship in East London*, Penguin, Harmondsworth.

SUBJECT INDEX

AUTHOR INDEX